# Lecture Notes in Mathematics

Edited by A. Dold and B. Eckmann

Series: Department of Mathematics,
University of Maryland, College Park
Adviser: L. Greenberg

492



# Infinitary Logic: In Memoriam Carol Karp

A Collection of Papers by Various Authors

Edited by D. W. Kueker

Springer-Verlag
Berlin · Heidelberg · New York 1975

**Editor**
Prof. David W. Kueker
Department of Mathematics
University of Maryland
College Park
Maryland 20705/USA

Library of Congress Cataloging in Publication Data

Main entry under title:

Infinitary logic.

(Lecture notes in mathematics ; 492)
1. Infinitary languages--Addresses, essays, lectures. 2. Model theory--Addresses, essays, lectures. I. Karp, Carol, 1926-1972. II. Kueker, David W., 1943- III. Series: Lecture notes in mathematics (Berlin) ; 492.
QA3.L28 no. 492 [QA9.37] 510'.8s [511'.3]
75-34464

AMS Subject Classifications (1970): 02B25, 02H10, 02H13

ISBN 3-540-07419-8 Springer-Verlag Berlin · Heidelberg · New York
ISBN 0-387-07419-8 Springer-Verlag New York · Heidelberg · Berlin



Printed in Germany

Offsetdruck: Julius Beltz, Hemsbach/Bergstr.

CAROL KARP

## CONTENTS



*Each part has its own complete table of contents*

## ACKNOWLEDGEMENTS

The origin of the present volume is explained in the Introduction. I wish to thank the editors of Springer's Perspectives in Mathematical Logic series for suggesting its preparation. E.G.K. López-Escobar has generously given advice and suggestions during the planning and editing of the book. I am also grateful to Pat Berg who carefully typed the entire volume.

David W. Kueker

# INTRODUCTION

BY

E.G.K. LÓPEZ-ESCOBAR

# INTRODUCTION

Carol Karp died on August 20, 1972 after a brave battle against cancer which had lasted for three long years[1]. To her, teaching had always been more than a duty, and even during her illness she taught all her classes in addition to carrying out her administrative tasks. Her research, too, was pushed forward with her usual determination, but unfortunately the planned new monograph on infinitary languages remained unfinished. Her early work was collected in her one published book, but she realized that it very much needed to be brought up to date.

Towards the end she was rather apprehensive that her doctoral students would not be able to complete their studies; but her fears were unfounded: they all now have their Ph.D. degrees. Her students are: R.J. Gauntt (Ph.D. 1969, now at California State College, Fullerton), J. Gregory (Ph.D. 1969, now at S.U.N.Y., Buffalo), J. Green (Ph.D. 1972, now at Rutgers, Camden) and E. Cunningham (Ph.D. 1974, now at St. Mary-of-the-Woods College, Indiana). To them as to us the memory of the conduct of her life, exceptional spirit and warm personality persists as a lasting inspiration.

## THE PLANNED MONOGRAPH

The new book was intended for the Springer series, Perspectives in Mathematical Logic. Carol Karp was in the habit of doing most of her own research work during the summer months. Unfortunately the necessary treatment in 1972 left her exhausted and vulnerable to virus infections, so that there was barely time for her to organize the material for the beginning chapters. The final chapter was to be based on her lectures Generalized Recursion

[1] **Her biography and bibliography follow this introduction.**

Theories at the Manchester Summer School (1969, unpublished), and from her notes it is apparent that she viewed it as the greatest contribution of the monograph. The gist of the programme was to show that the most natural way of generalizing recursion theory was through representability in formal theories with infinitely long formulas. The proposed outline of the whole book was as follows:

| | |
|---|---|
| CHAPTER I. | Partial isomorphisms. Type of structures. |
| CHAPTER II. | Infinitary formulas. |
| CHAPTER III. | $L_{\alpha\beta}$ $L_{\alpha\omega}$-model theory: consistency properties. |
| CHAPTER IV. | Admissible sets. |
| CHAPTER V. | $L_A$ - model theory: countable case, cofinality $\omega$ case. |
| CHAPTER VI. | Implicitly representable predicates. |

## THE PRESENT VOLUME AND CONTRIBUTORS

As the notes for the new monograph were so sparse, the editors of the Springer series decided that it would do more justice to her name if a separate book were written in her memory rather than commission anyone to complete her outline. Furthermore, in view of Carol Karp's strong interest in the development of mathematical logic at the University of Maryland, it was agreed that the volume be published under the Maryland section of the Springer Lecture Notes in Mathematics.

Since in the section headings of Chapter I, there were many references to the automorphism results of David Kueker and since he is also at the University of Maryland he has kindly agreed to be both editor and a contributor to the book.

The contents of Chapters III and IV and V had not been

subdivided into sections. However, from the notes available, it was clear that the results in Judy Green's thesis were to be an important part of those chapters. Green thus also agreed to be another of the contributors. Green's results concern languages $L_{\alpha\omega}$ (especially for $cf(\alpha) = \omega$) and their fragments, however Carol Karp had also planned to include results on the languages $L_{\alpha\alpha}$ ($cf(\alpha) = \omega$). At the time she agreed to write the new monograph Karp was not sure what exactly would be the results. She had obtained some independence results in her thesis through the use of general models. The general models were later changed to $\omega$-chains and the latter were applied to initiate a theory of models for the languages $L_{\alpha\alpha}$ ($cf(\alpha) = \omega$); some of the preliminary results were published in the Tarski Festschrifft. Karp had asked her student, Sr. Ellen Cunningham, to work out the theory of $\omega$-chains. Sr. Cunningham is contributing an account of her results.

The notes on Chapter VI were even more scarce. However, Karp did give lectures at Maryland in the Fall of 1970 on infinitary logic and recursion theory. Basically Karp wanted to return to Gödel's orginal proof-theoretic definition of recursive sets but of course using more liberal notions of proof so as to obtain generalizations of recursion theory. Now although the final formulation was to be proof-theoretic (and thus consistent with her general viewpoint) she was led to it by the model-theoretic definability criteria put forward by Kreisel in the 60's. The problem of the right formulation of the finiteness (or compactness) theorem for the infinitary languages gave rise to many interesting results. In 1968 Kreisel formulated a "generalized finiteness theorem" and asked if it held in every admissible set. Karp gave the problem to John Gregory who showed in 1969 that the answer was negative. The proof appears in Gregory's thesis; in order to make it available he has agreed to include the relevant sections of his thesis in this volume.

## A REVIEW OF KARP'S RECENT WORK

The first book on infinitary languages had had a very favorable reception, nevertheless, Karp was not completely satisfied with it. In the first place, the subject had progressed enormously in the last eight years; secondly, the style of the book was perhaps too austere. She was convinced that a freer style would be a great improvement; she also wanted to make sure that no librarian would dare catalogue the new book in the dead languages section (as apparently happened to the earlier work at a British university). This change in outlook went beyond her monographs. She always looked for a coherent view of logic and naturally as the subject developed what made good sense in 1957 need not do so in 1972. It is thus not surprising to find that her view on the role for infinitary languages changed during her lifetime.

At first, although she obviously enjoyed working with infinitely long formulae, she did not appear to have a very high opinion of them. For example in a 1960 research proposal to the National Science Foundation she states:

> *From the point of view of metamathematics,"formal" calculi based on languages with expressions of infinite length and having infinitely long formal proofs, are of no value. But in recent years, it has been noticed that algebraic results can come from the study of formal systems.....From this point of view, it is very profitable to consider infinitary formal calculi.*

It is probably fair to state that at the time when she was writing her thesis Karp considered herself to be principally an "algebraic logician". Her inclination towards algebra was never completely forgotten and she always seemed able to draw results concerning

Boolean algebras from her results about the infinitary languages. She had also obtained results about other structures, for example in the first chapter of her thesis there are some results about embedding conjunctive implicative algebras in Brouwerian algebras. In 1966 someone asked her about her results in implicative algebras; she replied stating precisely what were the results she had obtained and added the remark, *"For me, Boolean is best"*.

There are basically two ways of defining infinitely long formulae; either as transfinite well-ordered sequences of symbols or as unordered sets. The former is the more natural extension of the first-order predicate calculus but as technical complications arise with the operation of substitution most people choose the latter. However, Karp perservered with infinitely long formulae as transfinite sequences of symbols and worked out the required theory of transfinite concatenation (it appears in the second chapter of her North Holland book).

Once the nature of the infinitely long formulae has been settled the next problem is that of completeness. In the case of the infinitary languages $L_{\alpha\beta}$ (admitting conjunctions of length $< \alpha$ and quantification over sequences of individual variables of length $< \beta$) we have all the obvious extensions of the axioms and rules of inference of the first-order predicate calculus $L_{\omega\omega}$ to consider. In the case of $L_{\omega_1\omega}$ we do get a complete axiomatization. For the languages $L_{\alpha\beta}$ with $\alpha > \omega_1$ the situation was more problematic. On the one hand Karp had shown that the obvious extensions of the rules and axioms of $L_{\omega\omega}$ to $L_{\alpha\beta}$ $(\alpha > \omega_1)$ did not give complete axiomatizations. On the other hand she had shown that for cardinals $\alpha, \beta$ such that $\delta^{\varepsilon} < \alpha$ whenever $\delta < \alpha$ and $\varepsilon < \beta$ all that was required to obtain a complete axiomatization was to add Chang's distributive laws and the rule of $\gamma$-dependent choices for each

$\gamma < \alpha$. Of course, it is always possible to obtain a complete axiomatization by adding enough axioms, so the really fundamental question was what kind of complete axiomatizations should be considered (permitted). Dana Scott had considered definability conditions and showed that if $\alpha$ is a successor cardinal, then there is no complete axiomatization for $L_{\alpha\alpha}$ which is definable in $\langle H_\alpha, \in \rangle$ where $H_\alpha$ is the collection of sets hereditarily of power $< \alpha$. Karp preferred to concentrate on finding a suitable notion of effectiveness to deal with the problem of an ($\alpha$-) effective axiomatization for $L_{\alpha\beta}$.

In the early 60's Takeuti had introduced the concept of ordinal recursive functions, and so a possible way to tackle the problem of an effective axiomatization for $L_{\alpha\beta}$ was through ordinal recursiveness. One of Karp's research grants was precisely on that project. In a preliminary report she writes:

> *During the first year of the contract we classified the infinitary languages according to whether or not they were effectively axiomatizable, using as our criterion of effectiveness the notion of ordinal recursiveness.*

In order to be able to code the infinite formulae by ordinals, one has, in effect, to be able to code sequences of ordinals by ordinals. The latter cannot be done unless one assumes some further axioms of set theory, for example, the axiom of constructibility ($V = L$). Thus Karp's characterization of the infinitary languages (and Boolean algebraic results which she drew from it) depended on the hypothesis $V = L$. It was clear that she planned to eliminate the assumption of constructibility. In a footnote to an unpublished paper entitled *Applications of the theory of transfinite computability to infinitary formal systems* she states:

> *The axiom of constructibility was eliminated in the hierarchy of ordinal predicates by Lévy. The possibility of eliminating in this context as well, is*

*under investigation.*

This paper was never published in this form because she did succeed in eliminating the hypothesis $V = L$. In order to do so she had to introduce the primitive recursive set functions. Set-recursiveness not only avoided $V = L$, but also made the proofs much easier to understand. In addition, Karp used them to give a very nice proof of Gödel's consistency theorem of the continuum hypothesis; it appears in the Proceedings of the Summer School in Mathematical Logic, Leicester 1965 (published 1966).

During her stay in Leicester, Karp found out that Ronald Jensen had also discovered the primitive recursive set functions (in his research on the structure of the constructible universe). They combined their results, and produced an excellent paper *Primitive recursive set functions* (1968).

The discovery of the primitive recursive set functions, as well as the positive attitude towards infinite languages which prevailed at that time, had a marked effect on Carol Karp's own attitude to those languages. The infinitary languages were now not only useful for obtaining results about Boolean algebras; but were also useful in understanding other parts of mathematical logic; for example generalizations of recursive function theory. However her first attempt to use infinitary languages as an aid to understanding was not with respect to recursion theory, but with respect to set theory. The following is extracted from her 1965 proposal to the National Science Foundation entitled *Applications of infinitary logic to set theory.*

> *Axiomatic set theory is now in a critical stage with the methods of P.J. Cohen leading to more and more questions that the current systems do not settle. Part of the difficulty may be that axiomatic set theory, being a first-order theory, is ultimately an arithmetic of the natural numbers....A system* $ZF_{\Omega}$ *is described, an infinitary*

*analogue of Zermelo Fraenkel set theory, which is demonstrably stronger and retains such desirable characteristics as a complete underlying logic and formally representable proof predicate.*

It turned out that the infinitary languages $L_{\Omega\Omega}$ did not give any really new sets; apparently some new kinds of quantifiers are needed. Karp decided that it was premature to do any further work in that direction and that it would be advantageous to concentrate on other possible applications of infinitary languages.

Although Carol Karp appreciated recursive function theory, she disliked proofs which involved codings and systems of notations. In her work on infinitary set theory she noticed that infinitely long formulae sometimes allowed her to circumvent notations. She then tried treating the recursive set functions as functions representable in infinitary systems using implicit definitions. She discovered that by varying the logic in the system one could get a host of results about recursion theory and its extensions; furthermore it could be done without any ad hoc notations. Unfortunately, she only had time to work out some of the details for fragments of infinitary languages of the form $L_{\alpha\omega}$ (i.e., finite-quantifier infinitary languages). She presented some of the details at the Manchester meeting of the Summer School in Logic (1969). Regrettably Karp did not have time to write up a paper for that meeting. From some scattered note one gathers that her plans were roughly as follows:

Suppose that $A$ is an admissible set. For each $a \in A$, let $c_a$ be an individual constant symbol and then let $T_A$ be the set $\{(\forall v)(v \in c_a \longleftrightarrow \bigvee_{b \in a} v = c_b): a \in A\}$. Given an n-ary relation $R$ on $A$ and a finite formula $\varphi$ in the non-logical symbols $\in$, $R$, $S$, $c_a$, which is true in the structure $\langle A, \in, R, S, a\rangle$ where $a \in A$ and $S \subseteq A$ and a logic $L$ (i.e., a notion of consequence) then $\varphi$ *weakly implicitly represents* $R$ *on* $A$ *with respect to the logic* $L$

$$(\forall x_1 \ldots x_n \in A)(R(x_1,\ldots,x_n) \equiv T_A, \varphi \vdash_L Rc_{x_1} \ldots x_{x_n}).$$

R is *L-R.E. on* A iff there is a formula $\varphi$ such that $\varphi$ weakly implicitly represents R on A with respect to the logic L.

A logic L is said to be *admitted by* A iff the relation $\{\psi \in A : \psi$ is a formula and $\vdash_L \psi\}$ is L-R.E. on A and for all formulas $\theta \in A$:

$$T_A \vdash_L \theta \equiv (\exists s \in A)(s \subseteq T_A \ \& \ s \vdash_L \theta).$$

As an example let us consider the following three logics:

$L_1$ is the logic obtained by the natural extension of the rules and axioms of $L_{\omega\omega}$ to the infinitary formulae in A. Karp called $L_1$ the "Boolean logic".

$L_2$ is $L_1$ plus Chang's distributive laws restricted to A.

$L_3$ is obtained by taking the model theoretic notion of consequence, i.e. $\vdash_{L_3}$ is just $\models$.

In terms of the above logics Karp obtained these characterizations:

$L_1$ is admitted by $A \equiv A$ is admissible.

R is $L_1$-R.E. on $A \equiv R$ is $\Sigma_1$ on A.

$L_2$ is admitted by $A \equiv A$ is $\mathcal{P}$-admissible ($\mathcal{P}$ for power set).

R is $L_2$-R.E. on $A \equiv R$ is admissible in $\mathcal{P}$ on A.

$L_3$ is admitted by $A \equiv A$ is strongly admissible.

R is $L_3$-R.E. on $A \equiv R$ is s.i.i.d. on A.

The underlying method was to show that the condition that A admitted L gave enough closure conditions on A to have the beginnings of a theory of recursively enumerable predicates applied to the L-R.E. predicates; for example, the enumeration and fixed point theorems.

If in addition to L being admitted by A, L is complete with respect to deductions from L-R.E. on A sets of formulae, then L has the following compactness theorem: if S is an L-R.E. on A set of sentences, then S has a model iff every subset of S, which is in A, has a model. The Barwise compactness theorem and the Barwise/Karp cofinality $\omega$ compactness theorem are immediate corollaries.

Relative recursiveness, in infinitary recursion theory, is an area which is still wide open; even in metarecursion theory there is no universally accepted notion of meta-recursive in. There are also complications when one tries to extend the notion of L-R.E. The trouble is that when one takes a subclass $B \subseteq A$ as a given predicate it amounts to adding the diagram of B to the given logic. Let $L^{\#}$ be the resulting logic, then it can happen that although A admits L it may not admit $L^{\#}$ (for ordinary recursion theory this does not happen because the hereditarily finite sets admit all logics).

The problem of relative recursiveness was included in Karp's last research proposal, but unfortunately there are no records of her progress towards a coherent solution.

It was her research on the infinitely long formal proofs that led Karp to the concept of L-R.E. on A. However, it is clear that the actual structure of the proofs is irrelevant, for all that is ever used is the consequence relation. Thus, for the purpose of discussing extensions of recursion theory, it does not make much sense to dwell too much upon the axioms and rules of inference. Consistency properties are a natural way of getting all the benefits of completeness while, at the same time, avoiding formal proofs.

Karp, who always looked for simplifying methods, applied the consistency properties to extend some of her earlier results, see for example the abstract of her lecture *From countable to cofinality* $\omega$ *in infinitary logic* (1971).

Since the set of theorems of a formal calculus can be inductively defined, one suspects that many of her results on generalized recursion theory could also be obtained, without using tricky codings, from the general theory of inductive definitions. Had Karp been alive, I am sure she would have considered the exact relation between her notion of L-R.E. on $\Lambda$ and some of the notions which can be defined in terms of inductive definitions. However, even if it turns out that all that can be done with (fragments of) $L_{\alpha\omega}$ can also be done with inductive definitions, the approach through infinitary languages has, I believe, a greater pedagogic value.

## A PERSONAL NOTE

It was Carol Karp who invited me to come to Maryland in 1966. Our interests at that time were very similar, though they later diverged. Even so she was always extremely helpful and remained a wonderful friend and colleague. I am happy to record here my appreciation and gratitude, but I do it with special regret in the realization that our association was far too brief.

## BIOGRAPHY AND BIBLIOGRAPHY

*Carol Ruth Karp (née: van der Velde)*

*Born:* Ottawa County, Michigan, August 10, 1926

*Married:* Arthur L. Karp, 1952

*Education:*

1948 B.A., Manchester College, North Manchester, Indiana

1950 M.A., Michigan State University, East Lansing, Michigan

1959 Ph.D., University of Southern California, Los Angeles

*Professional Career:*

1953-54 Instructor, New Mexico State University, Las Cruces, New Mexico

1958-60 Instructor, University of Maryland, College Park, Maryland

1960-63 Assistant Professor, University of Maryland, College Park, Maryland

1963-66 Associate Professor, University of Maryland, College Park, Maryland

1966-72 Professor, University of Maryland, College Park, Maryland

*Professional Societies:*

1953-72 Member of the Association for Symbolic Logic

1956-72 Member of the American Mathematical Society

1968-72 Consulting Editor for the Journal of Symbolic Logic

1966-69 Representative of the Association of Symbolic Logic to the National Academy of Sciences and National Research Council.

*Ph.D. Thesis:*

1959 *Languages with expressions of infinite length.* University of Southern California, iv+183 pp., directed by Professor Leon Henkin. The thesis was divided into four chapters as follows:

Chapter

I. IMPLICATIVE MODELS AND CONJUNCTIVE IMPLICATIVE MODELS

1. Embedding Implicative Models in Conjunctive Implicative Models.

2. Embedding Conjunctive Implicative Models in Brouwerian Algebras.

II. A FORMULATION OF LANGUAGES WITH WFFS OF INFINITE LENGTH

1. Infinitary Expressions.

2. Description of Languages $L_{\alpha\beta}$.

3. Free and Bound Variables, Substitution, in $L_{\alpha\beta}$.

III. ALGEBRAICALLY COMPLETE FORMALIZATIONS FOR LANGUAGES WITH WFFS OF INFINITE LENGTH

1. Formal Systems $\mathcal{P}_{\alpha}(E)$, $\mathcal{F}_{\alpha\beta}(E)$.

2. An Algebraic Characterization of Provability in $\mathcal{P}_{\alpha}(E)$, $\mathcal{F}_{\alpha\beta}(E)$.

3. An Algebraic Characterization of Provability in Systems $\mathcal{P}_{\alpha}(E)$, $\mathcal{F}_{\alpha\beta}(E)$ with Negation.

IV. SEMANTICALLY COMPLETE FORMALIZATIONS FOR LANGUAGES WITH WFFS OF INFINITE LENGTH.

1. Semantically Complete Formalizations for Languages $P_{\alpha}$.

2. Formalizations for $P_{\alpha}$ and the Theory of $\alpha$-complete Boolean Algebras.

3. Semantically Complete Formalizations for Languages $F_{\alpha\beta}$.

*Invited Lectures:*

1964 American University Institute on History and Philosophy of Science and Mathematics, D.C.

1966 Summer School of Mathematics, Logic and Tenth Logic Colloquium, Leicester, England.

1966 Five-day Lecture Series, Hannover, West Germany.

1966 Three-day Lecture Series, Hamilton, Ontario.

1967 Set Theory Institute, U.C.L.A.

1969 Summer School and Colloquium in Mathematical Logic, Manchester, England.

1971 Annual meeting of the Association for Symbolic Logic, New York.

*Abstracts:*

1958 *Formalizations of propositional languages with Wffs of infinite length*, Notices of the American Mathematical Society, vol. 5, page 172.

1958 *Formalizations of functional languages with Wffs of infinite length*, Notices of the American Mathematical Society, vol. 5, page 172.

1958 *Formalisms for* $P_\alpha$, $F_{\alpha\beta}$ *and* $\alpha$*-complete Boolean algebras*, Notices of the American Mathematical Society, vol.5, page 173.

1958 *Split semantic models*, Notices of the American Mathematical Society, vol. 5, page 679.

1964 *Interpreting formal languages in directed systems and structures*, The Journal of Symbolic Logic, vol. 29, page 155.

1965 *Primitive recursive set functions: a formulation with applications to infinitary formal systems*, Abstracts of talks at the Logic Colloquium, Leicester, pp. 18-19.

1966 *Applications of recursive set functions to infinitary logic*, The Journal of Symbolic Logic, vol. 31, page 698.

1972 *From countable to cofinality* $\omega$ *in infinitary model theory*, the Journal of Symbolic Logic, vol. 37, pp. 430-431.

*Articles:*

1962 *Independence proofs in predicate logic with infinitely long expressions*, The Journal of Symbolic Logic, vol. 27, pp. 171-188.

1963 *A note on the representation of complete Boolean algebras*, Proceedings of the American Mathematical Society, vol. 14, pp. 705-707.

1965 *Finite quantifier equivalence*, article in The Theory of Models, Proceedings of the 1963 Symposium at Berkeley, North-Holland Publishing Co., pp. 407-412.

1967 *A proof of the relative consistency on the Continuum Hypothesis*, article in Sets, Models and Recursion Theory, Proceedings of the 1965 Colloquium at Leicester, North-Holland Publishing Company, pp. 1-32.

1968 *An algebraic proof of the Barwise compactness theorem*, article in The Syntax and Semantics of Infinitary Languages, Lecture Notes in Mathematics, vol. 72, pp. 89-95.

1971 *Primitive recursive set functions*, article, written in collaboration with R. Jensen, in Axiomatic Set Theory, Proceedings of Symposia in Pure Mathematics, vol. 13, Part 1, pp. 143-176.

1974 *Infinite-quantifier languages and* $\omega$*-chains of models*, article in Proceedings of the Tarski Symposium, Proceedings of Symposia in Pure Mathematics, vol. XXV, pp. 225-232.

*Book:*

1964 Languages with Expressions of Infinite Length, Studies in Logic, North-Holland Publishing Co., Amsterdam 1964, xix-183.

PART A

# BACK-AND-FORTH ARGUMENTS AND INFINITARY LOGICS

BY

DAVID W. KUEKER

# CONTENTS

## PART A

# INTRODUCTION

This part of the present volume corresponds to Chapters I and II of the outline of Carol Karp's proposed monograph given in López-Escobar's introduction. She had only completed a rough draft of Chapter I and notes for the contents of the other chapter. As work progressed it became clear that, contrary to our original expectations, Chapter I needed extensive rewriting. I have, however, followed her outline and basic plan very closely. Her procedure was to first study partial-isomorphisms (or back-and-forth mappings) fairly thoroughly, including results like Theorem I.1.4 to indicate the interest of such a study. Then infinitary logics are introduced and justified as the logics corresponding to these mappings. Finally, some model theory for these logics is developed, with emphasis on those results obtained using back-and-forth methods.

This part is therefore similar to Barwise's superlative survey article [1], the main differences in coverage being that we include infinite quantifier logics and omit the applications to abelian groups and remarks on foundational significance. The interested reader is in any case urged to read Barwise's paper.

Since this is an introduction to infinitary logics, we assume no previous familiarity with them, and most of the results presented will be known to those readers who are familiar with these logics. We do assume a reasonable acquaintance with first-order model theory, for which we refer to [6]. Our terminology and notation are fairly standard. By a language we understand a set of symbols for functions and relations (all with just finitely many places) and for individual constants. Models for such languages are denoted by German letters $\mathfrak{A}$ , $\mathfrak{B}$ and their universes by the corresponding roman capitals A, B.

If $\kappa$ is an infinite cardinal then $\kappa^+$ is the cardinal successor of $\kappa$ and $cf(\kappa)$ is the least cardinal $\gamma$ such that some $\gamma$-termed sequence of elements of $\kappa$ is cofinal in $\kappa$. If $\kappa$ and $\lambda$ are cardinals then

$$\kappa^{<\lambda} = \sum \{\kappa^{\beta} : \beta < \lambda\}.$$

A fact occasionally used in various forms is that if $|A| = \kappa$ then there are exactly $\kappa^{\lambda}$ sequences of elements of $A$ of length $\lambda$ and exactly $\kappa^{<\lambda}$ sequences of elements of $A$ of length less than $\lambda$.

The end of a proof is denoted by the symbol $\dashv$.

# CHAPTER I

## BACK-AND-FORTH ARGUMENTS AND PARTIAL ISOMORPHISM

1. Back-and-Forth One Element at a Time. What is probably the original back-and-forth argument appears in the familar proof of Cantor's theorem that any two countable dense linear orderings without endpoints are isomorphic. Briefly, this argument is as follows. Let $\mathfrak{A} = \langle A,<^{\mathfrak{A}}\rangle$ and $\mathfrak{B} = \langle B,<^{\mathfrak{B}}\rangle$ be countable dense linear orderings without end-points, and assume $A = \{a_n : n < \omega\}$ and $B = \{b_n : n < \omega\}$. We say that a finite sequence $\langle a'_0,\ldots,a'_k\rangle$ from $A$ is *similar* to a finite sequence $\langle b'_0,\ldots,b'_k\rangle$ from $B$ if they are ordered in the same way, that is, $a'_i <^{\mathfrak{A}} a'_j$ holds iff $b'_i <^{\mathfrak{B}} b'_j$ holds for all $i,j \le k$. Since both orderings are dense and have neither first nor last elements, it is easy to see that the following holds:

(*) if $\langle a'_0,\ldots,a'_k\rangle$ is similar to $\langle b'_0,\ldots,b'_k\rangle$ then for any $a'_{k+1} \in A$ there is some $b'_{k+1} \in B$ such that $\langle a'_0,\ldots,a'_{k+1}\rangle$ is similar to $\langle b'_0,\ldots,b'_{k+1}\rangle$, and conversely, given $b'_{k+1}$ we can find such an $a'_{k+1}$.

So $\langle a_0\rangle$ and $\langle c_0\rangle$ are similar for any $c_0 \in B$, by (*) there is some $d_0 \in A$ such that $\langle a_0,d_0\rangle$ is similar to $\langle c_0,b_0\rangle$, by (*) again there is some $c_1 \in B$ such that $\langle a_0,d_0,a_1\rangle$ is similar to $\langle c_0,b_0,c_1\rangle$, etc. By induction, then, we obtain elements $c_k \in B$ and $d_k \in A$ for all $k$ such that $\langle a_0,d_0,a_1,d_1,\ldots,a_k,d_k\rangle$ is similar to $\langle c_0,b_0,c_1,b_1,\ldots,c_k,b_k\rangle$ for every $k$. The mapping which sends $a_k$ to $c_k$ and $d_k$ to $b_k$ then defines an isomorphism of $\mathfrak{A}$ onto $\mathfrak{B}$.

Let us recast this argument slightly. Call a function $f$ from a finite subset of A into B a *partial isomorphism* of $\mathfrak{A}$ into $\mathfrak{B}$ if the domain of f is $\{a'_0, \ldots, a'_k\}$ and $\langle a'_0, \ldots, a'_k \rangle$ is similar to $\langle f(a'_0), \ldots, f(a'_k) \rangle$. Condition (*) then becomes

(**) if f is any partial isomorphism and $a \in A$ (or, $b \in B$) then there is some partial isomorphism g such that $f \subseteq g$ and $a \in \mathrm{dom}(g)$ (or, $b \in \mathrm{ran}(g)$).

Using (**) we can construct a chain $f_0 \subseteq f_1 \subseteq \ldots$ of partial isomorphisms such that $g = \bigcup_{n \in \omega} f_n$ maps A onto B, and is therefore an isomorphism of $\mathfrak{A}$ onto $\mathfrak{B}$.

It is clear that this method of showing that two countable models are isomorphic depends only on the existence of a family of "partial isomorphisms" satisfying condition (**). The following definition expresses this property for arbitrary models.

Definition. Let $\mathfrak{A}$ and $\mathfrak{B}$ be models for a language L. $\mathfrak{A}$ and $\mathfrak{B}$ are *partially isomorphic*, written $\mathfrak{A} \cong_2 \mathfrak{B}$, if there is a non-empty set I of isomorphisms of submodels of $\mathfrak{A}$ onto submodels of $\mathfrak{B}$ with the *back-and-forth property*:

for any $f \in I$ and $a \in A$ (or, $b \in B$) there is some $g \in I$ such that $f \subseteq g$ and $a \in \mathrm{dom}(g)$ (or, $b \in \mathrm{ran}(g)$).

We write $I: \mathfrak{A} \cong_2 \mathfrak{B}$ to indicate that I is a set of isomorphisms with the back-and-forth property. (The explanation of the subscript "2" is to be found in the generalizations in the next section.)

The above argument for Cantor's theorem now splits into two parts. First, since (*) did not depend on the countability of A or B, there is the following observation on dense linear orders.

1.1 THEOREM. *Any two dense linear orderings without end-points are partially isomorphic.*

Secondly, there is the following result which holds in complete generality.

1.2 THEOREM. *If* $\mathfrak{A}$ *and* $\mathfrak{B}$ *are countable and* $\mathfrak{A} \cong_2 \mathfrak{B}$ *then* $\mathfrak{A} \cong \mathfrak{B}$. *Furthermore, if* $I: \mathfrak{A} \cong_2 \mathfrak{B}$ *and* $f \in I$ *then there is an isomorphism* $g$ *of* $\mathfrak{A}$ *onto* $\mathfrak{B}$ *such that* $f \subseteq g$.

PROOF: Assume that $I: \mathfrak{A} \cong_2 \mathfrak{B}$ where $A = \{a_n : n \in \omega\}$ and $B = \{b_n : n \in \omega\}$, and let $f \in I$. By repeated applications of the back-and-forth property we define a chain $f = f_0 \subseteq f_1 \subseteq \ldots$ of elements of $I$ such that $a_k \in \mathrm{dom}(f_{2k+1})$ and $b_k \in \mathrm{ran}(f_{2k+2})$ for all $k$. Then $g = \bigcup_{n \in \omega} f_n$ maps $A$ onto $B$ and is an isomorphism since the union of a chain of isomorphisms is an isomorphism. ⊣

Remarks. (1) If $\mathfrak{A} \cong \mathfrak{B}$ then $\mathfrak{A} \cong_2 \mathfrak{B}$. In fact, if $f$ is an isomorphism of $\mathfrak{A}$ onto $\mathfrak{B}$ then $\{f\}: \mathfrak{A} \cong_2 \mathfrak{B}$.

(2) There are many other uses in the literature of back-and-forth arguments to show that any two countable models in a certain class are isomorphic. For example, the class of dense linear orderings with first and last elements, the class of atomless boolean algebras, the class of $\omega$-saturated models of a complete theory (see [6]), the class of abelian torsion groups with prescribed Ulm invariants (this is Ulm's Theorem, see [1]), and the class of models $\mathfrak{A} = \langle A, <^{\mathfrak{A}}, P^{\mathfrak{A}} \rangle$ where $\langle A, <^{\mathfrak{A}} \rangle$ is a dense linear ordering without endpoints and both $P^{\mathfrak{A}}$ and $A - P^{\mathfrak{A}}$ are dense subsets of $A$ (see [25]).

In many cases, as in all the examples given here, the same

argument shows that any two models in the class are partially isomorphic. This is not accidental, as we will see in Section II.3.

(3) $\cong_2$ is an equivalence relation. Checking transitivity requires some work but is basically straightforward.

(4) $\cong_2$ is preserved by direct products. If $I_1: \mathfrak{A}_1 \cong_2 \mathfrak{B}_1$ and $I_2: \mathfrak{A}_2 \cong_2 \mathfrak{B}_2$, let $I$ consist of all functions $f$ with domain contained in $A_1 \times A_2$ such that there are $f_1 \in I_1$ and $f_2 \in I_2$ such that $f(a_1,a_2) = \langle f_1(a_1), f_2(a_2)\rangle$ for all $\langle a_1,a_2\rangle$ in the domain of $f$. Then $I: \mathfrak{A}_1 \times \mathfrak{A}_2 \cong_2 \mathfrak{B}_1 \times \mathfrak{B}_2$. The same is true of infinite direct products, reduced direct products and many other algebraic operations (for more information see [2], [3], [8]).

(5) We could have required in the definition of partial isomorphism that all the functions in the set I have finitely generated submodels as domain and range. In fact, if $I: \mathfrak{A} \cong_2 \mathfrak{B}$ and if we let $I'$ be the set of all restrictions of functions in $I$ to finitely generated domains, then it is easily verified that $I': \mathfrak{A} \cong_2 \mathfrak{B}$.

(6) Notice that Theorem 1.2 holds assuming just that $\mathfrak{A}$ and $\mathfrak{B}$ are countably generated.

(7) Let $\mathfrak{A} \cong_2 \mathfrak{B}$, and define $(a_0,\ldots,a_{n-1}) \sim_2 (b_0,\ldots,b_{n-1})$ to hold iff $(\mathfrak{A},a_0,\ldots,a_{n-1}) \cong_2 (\mathfrak{B},b_0,\ldots,b_{n-1})$. Then $\sim_2$ has the following properties:

(*i*) $0 \sim_2 0$

(*ii*) if $(a_0,\ldots,a_{n-1}) \sim_2 (b_0,\ldots,b_{n-1})$ then for any $a_n \in A$ (or $b_n \in B$) there is some $b_n \in B$ (or, $a_n \in A$) with $(a_0,\ldots,a_n) \sim_2 (b_0,\ldots,b_n)$

(*iii*) if $(a_0,\ldots,a_{n-1}) \sim_2 (b_0,\ldots,b_{n-1})$ then they satisfy precisely the same atomic formulas.

Conversely, the existence of a relation $\sim$ satisfying (*i*), (*ii*) and (*iii*) implies that $\mathfrak{A} \cong_2 \mathfrak{B}$.

(8) The historically minded should note that Cantor's original proof is not a back-and-forth argument (see [4, p. 124]), but Hausdorff's proof ([10, p. 99]) does use such an argument. Hausdorff also gives the following interesting application (p. 455): there is a continuous 1 - 1 function mapping the reals onto the reals such that $h(0) = \pi$ and if $q$ is any non-zero rational then $h(q)$ is rational. First we obtain by Cantor's theorem a 1 - 1 order-preserving function $f$ mapping the set $A$ of rationals onto the set $B = A \cup \{\pi\}$ such that $f(0) = \pi$. The unique order-preserving extension of $f$ to the set of reals is then the desired $h$. Further such applications are given by Skolem [25].

If $\mathfrak{A} \cong_2 \mathfrak{B}$ then any countable $\mathfrak{A}_0 \subseteq \mathfrak{A}$ is isomorphic to a submodel of $\mathfrak{B}$ (simply build a chain of partial isomorphisms such that the union is defined on all of $A_0$). In fact, rather stronger statements are true (see [16]). The following result is an instructive example.

1.3 THEOREM. *Assume that* $\mathfrak{A} \cong_2 \mathfrak{B}$ *and let* $a_0, a_1 \in A$ *and* $b_0, b_1 \in B$ *be such that* $a_0 \sim_2 b_0$ *and* $a_1 \sim_2 b_1$. *Then there are countable* $\mathfrak{A}_0 \subseteq \mathfrak{A}$ *and* $\mathfrak{B}_0 \subseteq \mathfrak{B}$ *such that*

$$(\mathfrak{A}_0, a_0) \cong (\mathfrak{B}_0, b_0) \quad \textit{and} \quad (\mathfrak{A}_0, a_1) \cong (\mathfrak{B}_0, b_1).$$

PROOF: Let $I: (\mathfrak{A}, a_0) \cong_2 (\mathfrak{B}, b_0)$ and $J: (\mathfrak{A}, a_1) \cong_2 (\mathfrak{B}, b_1)$. By Remark (5) above we may assume all functions in $I$ and $J$ have finitely generated domain and range. Pick any $f_0 \in I$ with

$a_0, a_1 \in \text{dom}(f_0)$. By a finite number of applications of the back-and-forth property, there is some $g_0 \in J$ with $\text{dom}(f_0) \subseteq \text{dom}(g_0)$ and $\text{ran}(f_0) \subseteq \text{ran}(g_0)$. Similarly, there is some $f_1 \in I$ with $f_0 \subseteq f_1$, $\text{dom}(g_0) \subseteq \text{dom}(f_1)$ and $\text{ran}(g_0) \subseteq \text{ran}(f_1)$, etc. In this way we get chains $\{f_n\}_{n\in\omega}$ from $I$ and $\{g_n\}_{n\in\omega}$ from $J$ such that $f = \bigcup_{n\in\omega} f_n$ and $g = \bigcup_{n\in\omega} g_n$ have the same domain and range, say $\mathfrak{A}_0$ and $\mathfrak{B}_0$. Then $f$ is an isomorphism of $(\mathfrak{A}_0, a_0)$ onto $(\mathfrak{B}_0, b_0)$ and $g$ is an isomorphism of $(\mathfrak{A}_0, a_1)$ onto $(\mathfrak{B}_0, b_1)$ as desired. ⊣

As an easy consequence we show that well-orderings are determined up to isomorphism by partial isomorphism.

COROLLARY. *If* $\mathfrak{A} = \langle A, <^{\mathfrak{A}} \rangle$ *is a well-ordering and* $\mathfrak{A} \simeq_2 \mathfrak{B}$ *then* $\mathfrak{A} \cong \mathfrak{B}$.

PROOF: We will show that for any $a \in A$ there is exactly one $b \in B$ such that $a \sim_2 b$. That there is at least one such $b$ is clear, so assume there were two different elements with this property, say $b_1$ and $b_2$. Then by Theorem 1.3 there are countable $\mathfrak{A}_0 \subseteq \mathfrak{A}$ and $\mathfrak{B}_0 \subseteq \mathfrak{B}$ such that $(\mathfrak{A}_0, a) \cong (\mathfrak{B}_0, b_1)$ and $(\mathfrak{A}_0, a) \cong (\mathfrak{B}_0, b_2)$. But this implies that there are two different isomorphisms of $\mathfrak{A}_0$ onto $\mathfrak{B}_0$, which is impossible for well-orderings. So we can define $h$ on $A$ by $h(a)$ = the unique $b \in B$ with $a \sim_2 b$. This is then easily verified to be an isomorphism of $\mathfrak{A}$ into $\mathfrak{B}$; it is onto since every $b \in B$ satisfies $a \sim_2 b$ for some $a \in A$. ⊣

The relation of partial isomorphism divides the countable models into two natural classes, those which are determined up to isomorphism by partial isomorphism and those which are partially isomorphic to non-isomorphic models. By Theorem 1.2 the models in the second category are partially isomorphic to uncountable models. Such models have certain interesting properties. For example, they are isomorphic to proper submodels of themselves (this is left as an exercise), and they have $2^\omega$ automorphisms. This latter fact is derived from the following theorem.

1.4 THEOREM (Kueker [15]). *Let* $\mathfrak{A}$ *be countable. Then the following are equivalent:*

(*i*) $\mathfrak{A}$ *has* $2^\omega$ *automorphisms.*

(*ii*) $\mathfrak{A}$ *has uncountably many automorphisms.*

(*iii*) *For any finite sequence* $\vec{a}$ *of elements of* A, $(\mathfrak{A},\vec{a})$ *has a non-trivial automorphism.*

Before proving the theorem, we derive the corollary we are interested in.

COROLLARY (Kueker [15]). *If* $\mathfrak{A}$ *is countable and* $\mathfrak{A} \cong_2 \mathfrak{B}$ *for some uncountable* $\mathfrak{B}$, *then* $\mathfrak{A}$ *has* $2^\omega$ *automorphisms.*

PROOF: We will show that condition (*iii*) of Theorem 1.4 holds. Let $\vec{a}$ be a finite sequence from A. There is some sequence $\vec{b}$ from B such that $\vec{a} \sim_2 \vec{b}$. For each $d \in B$ there is some $c \in A$ with $(\vec{a},c) \sim_2 (\vec{b},d)$. Since A is countable and B is uncountable there is some $c_1 \in A$ and $d_1,d_2 \in B$ such that $d_1 \neq d_2$ but $(\vec{a},c_1) \sim_2 (\vec{b},d_1)$ and $(\vec{a},c_1) \sim_2 (\vec{b},d_2)$. Let $c_2 \in A$ be such that $(\vec{a},c_1,c_2) \sim_2 (\vec{b},d_1,d_2)$. It follows that $c_1 \neq c_2$. But $(\vec{a},c_2) \sim_2 (\vec{b},d_2)$ and $(\vec{b},d_2) \sim_2 (\vec{a},c_1)$, hence $(\mathfrak{A},\vec{a},c_2) \cong_2 (\mathfrak{A},\vec{a},c_1)$; that is, $(\mathfrak{A},\vec{a})$ has an automorphism mapping $c_2$ to $c_1$. ⊣

PROOF OF THEOREM 1.4: The implications (*i*) $\Longrightarrow$ (*ii*) $\Longrightarrow$ (*iii*) are clear. We show (*iii*) $\Longrightarrow$ (*i*) as follows.

Assume that for any finite sequence $\vec{a}$ of elements of A,

$(\mathfrak{A},\vec{a})$ has a non-trivial automorphism.

We write $\vec{a} \sim \vec{a}'$ to mean that $(\mathfrak{A},\vec{a}) \cong (\mathfrak{A},\vec{a}')$.

We first show that if $\vec{a} \sim \vec{a}'$ then there are $b,c,c' \in A$ such that $c \neq c'$ but $(\vec{a},b) \sim (\vec{a}',c)$ and $(\vec{a},b) \sim (\vec{a}',c')$. To see this, let $g$ be a non-trivial automorphism of $(\mathfrak{A},\vec{a})$ and let $b$ be such that $g(b) \neq b$. Then there are $c,c' \in A$ such that

$$(\vec{a},b,g(b)) \sim (\vec{a}',c,c').$$

This implies that $c \neq c'$, and also $(\vec{a},b) \sim (\vec{a}',c)$ and $(\vec{a},g(b)) \sim (\vec{a}',c')$, hence $(\vec{a},b) \sim (\vec{a}',c')$ since $(\vec{a},b) \sim (\vec{a},g(b))$.

Now let $S$ be the set of all functions $s$ with domain $\omega$ and range contained in $\{0,1\}$. If $s \in S$ and $n \in \omega$ then $s|n$ is the restriction of $s$ to domain $\{0,\ldots,n-1\}$. We will define finite partial mappings $h_{s|n}$ of $A$ into $A$ with the following properties:

(1) if the domain of $h_{s|n}$ is $\{c_0,\ldots,c_k\}$ then $(c_0,\ldots,c_k) \sim (h_{s|n}(c_0),\ldots,h_{s|n}(c_k))$;

(2) $a_n$ is in the domain and range of $h_{s|(n+1)}$;

(3) if $s|n = s'|n$ but $s(n) \neq s'(n)$ then $h_{s|(n+1)}(e) \neq h_{s'|(n+1)}(e)$ for some $e$;

(4) $h_{s|n} \subseteq h_{s|(n+1)}$ for all $n$.

Once these are defined we can define $h_s = \bigcup_{n\in\omega} h_{s|n}$ for each $s \in S$. Then each $h_s$ is well-defined by (4), maps $A$ onto $A$ by (2), and is an automorphism of $\mathfrak{A}$ by (1). Further, if $s \neq s'$ then (3) guarantees us that $h_s(e) \neq h_{s'}(e)$ for some $e$. Therefore the $h_s$'s form a collection of $|S| = 2^\omega$ different automorphisms of $\mathfrak{A}$ as desired.

The mappings $h_{s|n}$ are defined by induction on $n$. First, $h_{s|0} = 0$. Next assume we have defined $h_{s|n}$ with domain $\{c_0,\ldots,c_k\}$. To begin with there are $d,d' \in A$ such that

$$(c_0,\ldots,c_k,a_n,d) \sim (h_{s|n}(c_0),\ldots,h_{s|n}(c_k),d',a_n).$$

By our earlier remark there are $e,e',e'' \in A$ with $e' \neq e''$ such that

$$(c_0,\ldots,c_k,a_n,d,e) \sim (h_{s|n}(c_0),\ldots,h_{s|n}(c_k),d',a_n,e')$$

and

$$(c_0,\ldots,c_k,a_n,d,e) \sim (h_{s|n}(c_0),\ldots,h_{s|n}(c_k),d',a_n,e'').$$

Now just define $h_{s|(n+1)}$ to have domain $\{c_0,\ldots,c_k,a_n,d,e\}$ and extending $h_{s|n}$ so that

$$h_{s|(n+1)}(a_n) = d',$$

$$h_{s|(n+1)}(d) = a_n,$$

and

$$h_{s|(n+1)}(e) = e' \quad \text{if } s(n) = 0,$$

$$h_{s|(n+1)}(e) = e'' \quad \text{if } s(n) = 1.$$

It is then clear that $h_{s|(n+1)}$ has the desired properties (1)-(4).

It is easy to find countable models with $2^\omega$ automorphisms which are not partially isomorphic to uncountable models. In [19] Makkai gave an interesting characterization of the countable models which are partially isomorphic to uncountable models and applied it to linear orderings.

The following theorem is suggested by the Chang-Makkai theorem of finitary logic [6, p. 255]. Its proof is similar to that of

Theorem 1.4, and is therefore left as an exercise.

1.5 THEOREM (Kueker [15], Reyes [22]). *Let* $\mathfrak{A}$ *be countable and let* $P \subseteq A$. *Then the following are equivalent:*

(*i*) $|\{Q : (\mathfrak{A},P) \cong (\mathfrak{A},Q)\}| = 2^\omega$

(*ii*) $|\{Q : (\mathfrak{A},P) \cong (\mathfrak{A},Q)\}| > \omega$

(*iii*) *For any finite sequence* $\vec{a}$ *of elements of* $A$ *there are* $b \in P$ *and* $c \in A - P$ *such that*

$$(\mathfrak{A},\vec{a},b) \cong (\mathfrak{A},\vec{a},c).$$

There is another equivalent condition that can be added to Theorem 1.4, with a proof similar to that showing (*iii*) implies (*i*). Makkai has pointed out that it also can be derived from Theorem 1.5, which is what we will do here.

COROLLARY (Kueker [15]). *For any countable* $\mathfrak{A}$ *the following condition is equivalent to those listed in Theorem* 1.4*:*

(*iv*) *There is some* $P \subseteq A$ *such that*

$$|\{Q : (\mathfrak{A},P) \cong (\mathfrak{A},Q)\}| = 2^\omega.$$

PROOF: It is clear that (*iv*) implies the other conditions of 1.4. We will assume (*iii*) of 1.4 and show (*iv*) by showing that (*iii*) of 1.5 holds. Let $A = \{a_n : n \in \omega\}$. We inductively define $c_n, d_n \in A$ such that $c_n \neq d_n$ but $(\vec{s}_n, c_n) \sim_2 (\vec{s}_n, d_n)$ for all $n$, where

$$\vec{s}_n = (a_0,\ldots,a_{n-1},c_0,\ldots,c_{n-1},d_0,\ldots,d_{n-1}).$$

This is easy using (*iii*) of 1.4. Now define $P = \{c_n : n \in \omega\}$. Then clearly $d_n \notin P$ for all $n$. Then for any $n$, there is an automorphism of $(\mathfrak{A}, a_0, \ldots, a_{n-1})$ which maps $c_n$ to $d_n$, and so (*iii*) of 1.5 holds. ⊣

The results of this section hopefully will serve as an indication of the mathematical interest of partial isomorphism. For its foundational significance the reader is referred to [1].

2. Back-and-Forth Fewer than $\kappa$ Elements at a Time. In this section we present the generalization of back-and-forth mappings to higher cardinalities.

Let $\kappa$ be any cardinal greater than 1.

Definition. $\mathfrak{A}$ and $\mathfrak{B}$ are $\kappa$-*partially-isomorphic*, written $\mathfrak{A} \cong_\kappa \mathfrak{B}$, if there is a non-empty set $I$ of isomorphisms of submodels of $\mathfrak{A}$ onto submodels of $\mathfrak{B}$ with the $\kappa$-*back-and-forth property*:

for any $f \in I$ and $C \subseteq A$ with $|C| < \kappa$ (or, $D \subseteq B$ with $|D| < \kappa$) there is some $g \in I$ such that $f \subseteq g$ and $C \subseteq \mathrm{dom}(g)$ (or, $D \subseteq \mathrm{ran}(g)$).

We write $I: \mathfrak{A} \cong_\kappa \mathfrak{B}$ to indicate that $I$ is a set of isomorphisms with the $\kappa$-back-and-forth property.

Remarks. (1) When $\kappa = 2$, this is exactly the definition of partially isomorphic of Section 1.1.

(2) If $\mathfrak{A} \cong \mathfrak{B}$ then $\mathfrak{A} \cong_\kappa \mathfrak{B}$ for all $\kappa$, and if $\kappa < \lambda$ then $\mathfrak{A} \cong_\lambda \mathfrak{B}$ implies $\mathfrak{A} \cong_\kappa \mathfrak{B}$.

(3) It is easily seen that $\mathfrak{A} \cong_2 \mathfrak{B}$ implies $\mathfrak{A} \cong_\omega \mathfrak{B}$, so being partially isomorphic is the same as being $\omega$-partially isomorphic.

(4) $\cong_\kappa$ is an equivalence relation; once again, transitivity requires some little argument. Also, analogues of remarks (4), (5),

(6) and (7) of Section 1 hold, the exact formulation of which is left to the reader.

(5) If $|A| < \kappa$ and $I: \mathfrak{A} \cong_\kappa \mathfrak{B}$ then $\mathfrak{A} \cong \mathfrak{B}$, since there is some $f \in I$ with $A = \mathrm{dom}(f)$, and if $f$ did not map onto $B$ we could choose $b \in B - \mathrm{ran}(f)$ and find $g \in I$ with $f \subseteq g$ and $b \in \mathrm{ran}(g)$, which is impossible.

One often has stronger characterization results than that of Remark 5. As an example consider $\langle Re,<\rangle$, the reals under their natural ordering. We will show that $\langle Re,<\rangle \cong_{\omega_1} \mathfrak{A}$ implies that $\langle Re,<\rangle \cong \mathfrak{A}$. Let $I$ be such that $I: \langle Re,<\rangle \cong_{\omega_1} \mathfrak{A}$. Then there is some $f \in I$ such that the domain of $f$ includes the set $Q$ of rationals. For any real $r$ there is some $g \in I$ with $f \subseteq g$ and $r \in \mathrm{dom}(g)$; we claim that if $g' \in I$ is also such that $f \subseteq g'$ and $r \in \mathrm{dom}(g')$ then $g(r) = g'(r)$. Knowing that, we can define $h$ on $Re$ by $h(r) =$ the unique value $g(r)$ for any $g \in I$ with $f \subseteq g$ and $r \in \mathrm{dom}(g)$, and $h$ is easily verified to be an isomorphism of $\langle Re,<\rangle$ onto $\mathfrak{A}$. So, let such $g, g'$ be given and assume that $g(r) \neq g'(r)$. Let $h \in I$ be such that $g \subseteq h$ and $g'(r) \in \mathrm{ran}(h)$, say $h(r') = g'(r)$. Then $r' \neq r$, so there is some rational $q$ between them, say $r < q < r'$. Then, in $\mathfrak{A}$ we have: $f(q) = h(q) < h(r')$ and $g'(r) < g'(q) = f(q)$; but since $g'(r) = h(r')$ this is a contradiction. Hence $g(r) = g'(r)$, which completes the argument.

It must be emphasized, however, that $\kappa$-partial isomorphism does not play the same role in back-and-forth arguments with models of power $\kappa$ as 2-partial isomorphism does with countable models. This is due to the failure of the generalization of Theorem 1.2 to uncountable $\kappa$ -- $\kappa$-partially isomorphic models of power $\kappa$ need not be isomorphic. The trouble is that the $\kappa$-back-and-forth property provides no machinery for continuing an induction past limit ordinals.

For example, let $\mathfrak{A}$ and $\mathfrak{B}$ have cardinality $\omega_1$ and let $I: \mathfrak{A} \cong_{\omega_1} \mathfrak{B}$. We naively try to construct an isomorphism of $\mathfrak{A}$ onto $\mathfrak{B}$ as follows. Pick $f_0 \in I$, next pick $f_1 \in I$ such that $f_0 \subseteq f_1$ and $\mathrm{dom}(f_1)$ contains some chosen countable subset of $A$, then pick $f_2 \in I$ such that $f_1 \subseteq f_2$ and $\mathrm{ran}(f_2)$ contains some particular countable subset of $B$, and continue. After $\omega$ steps we have a chain $\{f_n\}_{n\in\omega}$ of elements of $I$, and their union, $g$ is an isomorphism of a submodel of $\mathfrak{A}$ onto a submodel of $\mathfrak{B}$. But we do not know that $g \in I$ so we do not know that we can extend $g$ to isomorphisms between larger and larger pieces of $\mathfrak{A}$ and $\mathfrak{B}$.

The following definition strengthens the previous one by adding a union-of-chains condition which allows us to actually carry out the above naive construction.

Definition. $\mathfrak{A}$ and $\mathfrak{B}$ are *strongly $\kappa$-partially isomorphic*, written $\mathfrak{A} \cong^s_\kappa \mathfrak{B}$, if there is a set $I$ such that $I: \mathfrak{A} \cong_\kappa \mathfrak{B}$ and $I$ is closed under unions of chains of length less than $\mathrm{cf}(\kappa)$.

It is now a straightforward matter to check that the following generalization of Theorem 1.2 holds.

2.1 THEOREM. *If $\mathfrak{A}$ and $\mathfrak{B}$ are models of cardinality $\kappa$, then $\mathfrak{A} \cong^s_\kappa \mathfrak{B}$ iff $\mathfrak{A} \cong \mathfrak{B}$.*

If $\mathrm{cf}(\kappa) = \omega$, then $\cong^s_\kappa$ is the same as $\cong_\kappa$, so we obtain the following result first noted by Chang.

COROLLARY (Chang [5]). *If $\mathfrak{A}$ and $\mathfrak{B}$ have cardinality $\kappa$ and $\mathrm{cf}(\kappa) = \omega$ then $\mathfrak{A} \cong_\kappa \mathfrak{B}$ iff $\mathfrak{A} \cong \mathfrak{B}$.*

There are examples to show that $\cong^s_\kappa$ is strictly stronger than

$\cong_\kappa$ for all $\kappa$ of uncountable cofinality; in fact there are models $\mathfrak{A}$ and $\mathfrak{B}$ of cardinality $\kappa$ such that $\mathfrak{A} \cong_\kappa \mathfrak{B}$ but $\mathfrak{A} \not\cong \mathfrak{B}$. For regular $\kappa$ these were found by M. Morley (unpublished); the case of singular $\kappa$ with uncountable cofinality was recently settled by W. Tait [26].

Several familiar uncountable back-and-forth arguments provide examples of strong $\kappa$-partial isomorphism.

(1) A model $\mathfrak{A} = \langle A, <^{\mathfrak{A}} \rangle$ of dense linear order without end-points is an $\eta_1$-set if whenever $C_0$ and $C_1$ are countable subsets of $A$ such that every element of $C_0$ precedes every element of $C_1$ in the ordering, then there are $a_1, a_2, a_3 \in A$ such that

$$a_1 <^{\mathfrak{A}} c_0 <^{\mathfrak{A}} a_2 <^{\mathfrak{A}} c_1 <^{\mathfrak{A}} a_3$$

holds for all $c_0 \in C_0$ and $c_1 \in C_1$. Hausdorff [10] showed that any two $\eta_1$-sets of power $\omega_1$ are isomorphic. Essentially the same argument shows that any two $\eta_1$-sets are strongly $\omega_1$-partially isomorphic; simply take $I$ to be the set of all isomorphisms between countable submodels of the two $\eta_1$-sets.

(2) Elementarily equivalent $\kappa$-saturated models of power $\kappa$ are isomorphic (see [6]). Here too, essentially the same argument shows that any two $\kappa$-saturated elementarily equivalent models are strongly $\kappa$-partially isomorphic (compare Corollary 2.4 of [13]). This implies, for example, that any two uncountable models of a countable $\omega_1$-categorical theory of finitary logic are strongly $\omega_1$-partially isomorphic (since they are $\omega_1$-saturated; see [6]).

(3) If $\mathfrak{A}$ and $\mathfrak{B}$ are both $\kappa$-homogeneous and $\mathfrak{A} \cong_2 \mathfrak{B}$ then $\mathfrak{A} \cong^s_\kappa \mathfrak{B}$ (compare Corollary 2.5 of [13]).

Despite Theorem 2.1 and the above examples, $\cong^s_\kappa$ has not been studied much, and we will not do more with it in this paper. Part of the reason for this, looking ahead to the next chapter, is that $\cong_\kappa$ corresponds to equivalence in the logic $L_{\infty\kappa}$, whereas $\cong^s_\kappa$ is not known to correspond to equivalence in any logic. Therefore results concerning $\cong_\kappa$ turn into definability and logical characterizability theorems, while results on $\cong^s_\kappa$ remain combinatorial. Furthermore, it is not even clear if $\cong^s_\kappa$ is transitive! This should serve to indicate the difficulties inherent in strong $\kappa$-partial isomorphism.

The corollary to Theorem 2.1 and the comments following it indicate that, at least as far as $\cong_\kappa$ and the standard infinitary logics are concerned, better results are forthcoming for $\kappa$ of countable cofinality than in general. In fact this phenomenon will be repeatedly encountered in the contributions of J. Green and E. Cunningham to the present volume. But even then everything does not go smoothly. Difficulties arise which led to Karp to develop a theory of "non-standard" models, called *chain models*, which are discussed in Cunningham's paper.

As a good illustration, consider the cofinality $\omega$ extension of the theorem on automorphisms, Theorem 1.4. If $cf(\kappa) = \omega$ and $\mathfrak{A}$ has power $\kappa$ then what one would expect is that $\mathfrak{A}$ has more than $\kappa$ automorphisms iff for any sequence $\vec{a}$ of fewer than $\kappa$ elements of $A$ $(\mathfrak{A},\vec{a})$ has a proper automorphism. Indeed, we will see that the implication from right to left is correct. But the other direction, which was the easy direction for $\kappa = \omega$, fails in general. We are thus stuck with a one-way theorem. Note that if $cf(\kappa) = \omega$ then $\kappa^\omega > \kappa$, and in fact $\kappa^\omega = 2^\kappa$ when $\kappa$ is also a strong limit (that is, if $\alpha < \kappa$ implies $2^\alpha < \kappa$).

2.2 THEOREM (Kueker [15]). *Let* $cf(\kappa) = \omega$ *and let* $\mathfrak{A}$ *be a model of power* $\kappa$. *Then* $\mathfrak{A}$ *has at least* $\kappa^{\omega}$ *automorphisms provided that* $(\mathfrak{A},\vec{a})$ *has a proper automorphism for every sequence* $\vec{a}$ *of fewer than* $\kappa$ *elements of* A.

PROOF: The hypothesis in fact implies that if $\vec{a}$ is a sequence of fewer than $\kappa$ elements of $A$ and if $\lambda < \kappa$ then there are $d_\xi \in A$, $\xi < \lambda$, such that $(\mathfrak{A},\vec{a},d_\xi)_{\xi<\mu}$ has an automorphism moving $d_\mu$, for each $\mu < \lambda$. Let $\vec{d}$ be such a sequence $\langle d_\xi : \xi < \lambda\rangle$. Then whenever $\vec{b}$ is such that $(\mathfrak{A},\vec{a}) \cong (\mathfrak{A},\vec{b})$ we can find $\lambda$ different sequences $\vec{e}^{\nu}$, $\nu < \lambda$, such that $(\mathfrak{A},\vec{a},\vec{d}) \cong (\mathfrak{A},\vec{b},\vec{e}^{\nu})$ for all $\nu < \lambda$. In particular there are at least $\lambda$ different automorphisms moving $\vec{a}$ to $\vec{b}$.

Since $cf(\kappa) = \omega$ there are cardinals $\kappa_n < \kappa$, $n \in \omega$, such that $\kappa = \bigcup_{n<\omega} \kappa_n$, and $A = \bigcup_{n<\omega} A_n$ where $|A_n| = \kappa_n$ for each n. Let $S = \times_{n<\omega} \kappa_n$ be the set of all functions s with domain $\omega$ and such that $s(n) \in \kappa_n$ for all $n \in \omega$.

We can now, using the above remark, imitate the proof for $\kappa = \omega$ and define partial functions $h_{s|n}$ with domain having cardinality less than $\kappa$ such that:

(1) if $\mathrm{dom}(h_{s|n}) = \{c_\xi : \xi < \lambda\}$ then

$$(\mathfrak{A},c_\xi)_{\xi<\lambda} \cong (\mathfrak{A},h_{s|n}(c_\xi))_{\xi<\lambda};$$

(2) $A_n$ is contained in the domain and the range of $h_{s|(n+1)}$;

(3) if $s|n = s'|n$ but $s(n) \neq s'(n)$ then there is some d such that $h_{s|(n+1)}(d) \neq h_{s'|(n+1)}(d)$;

(4) $h_{s|n} \subseteq h_{s|(n+1)}$.

The details are left as an exercise.

It then follows that the collection of all $h_s = \bigcup_{n<\omega} h_{s|n}$ for $s \in S$ is a set of distinct automorphisms of $\mathfrak{A}$ of cardinality

$$|S| = \prod_{n<\omega} \kappa_n = \kappa^\omega .$$

Although Theorem 2.2 is only a one-way theorem, it is strong enough to yield the following generalization of the corollary to 1.4.

COROLLARY (Kueker [15]). *Let* $\mathfrak{A}$ *have power* $\kappa$ *where* $\mathrm{cf}(\kappa) = \omega$. *If* $\mathfrak{A} \cong_\kappa \mathfrak{B}$ *for some* $\mathfrak{B}$ *of power greater than* $\kappa$, *then* $\mathfrak{A}$ *has at least* $\kappa^\omega$ *automorphisms.*

Let us now show that the converse to 2.2 in fact does not hold. Indeed, the following is true: for any $\kappa$ cofinal with $\omega$ there is a model $\mathfrak{A}$, for a language with at most $\kappa$ symbols, with $\kappa^\omega$ automorphisms but such that for some countable sequence $\vec{a}$, $(\mathfrak{A},\vec{a})$ has no proper automorphisms.

First define a model $\mathfrak{B}$ of power $\kappa$ with $\kappa$ automorphisms but such that $(\mathfrak{B},b_0)$ has no proper automorphisms for some $b_0 \in B$. This is left as an exercise. Now let $\mathfrak{A}$ be the union of $\omega$ disjoint copies of $\mathfrak{B}$ with $\omega$ new unary predicates interpreted as the universes of these copies. Then $\mathfrak{A}$ has $\kappa^\omega$ automorphisms, since any $\omega$-sequence of automorphisms of $\mathfrak{B}$ can be pieced together to yield an automorphism of $\mathfrak{A}$. But if $\vec{a}$ is the countable sequence containing the element $b_0$ from each copy of $\mathfrak{B}$, then $(\mathfrak{A},\vec{a})$ has no proper automorphisms.

Karp noticed, however, that in the context of her chain-models one naturally obtains a biconditional version of 2.2. The notion of automorphism for such models is *bounded* automorphism, and her result states that for a strong limit cardinal $\kappa$ cofinal with $\omega$, a chain-model of (strict) power $\kappa$ has $2^\kappa$ bounded automorphisms iff a condition similar to that in 2.2 holds. The relevant definitions and precise statement are in Cunningham's paper in this volume.

Examples can be found showing that Theorem 2.2 is false for $\kappa$

of uncountable cofinality. However, the corollary to 2.2 can be generalized using strong $\kappa$-partial isomorphism. We omit the proof, since it will be included in a future paper of the author.

2.3 THEOREM. *Let* $\mathfrak{A}$ *have cardinality* $\kappa$ *and assume that* $\mathfrak{A} \cong_{\kappa}^{s} \mathfrak{B}$ *for some* $\mathfrak{B}$ *of cardinality greater than* $\kappa$. *Then* $\mathfrak{A}$ *has (at least)* $\kappa^{cf(\kappa)}$ *automorphisms.*

3. Ranked Partial Isomorphisms. Once again we let $\kappa$ be any cardinal greater than 1. The relation $\cong_{\kappa}$ is quite strong, and for some uses we need weaker relations approximating $\cong_{\kappa}$. Here we introduce, for each ordinal $\alpha$, the relation $\cong_{\kappa}^{\alpha}$. These have the property that $\mathfrak{A} \cong_{\kappa} \mathfrak{B}$ if and only if $\mathfrak{A} \cong_{\kappa}^{\alpha} \mathfrak{B}$ for all $\alpha$, and are also closely connected with the infinitary logics to be introduced in the next chapter.

Definition. $\mathfrak{A}$ and $\mathfrak{B}$ are *$\kappa$-partially isomorphic up to* $\alpha$, written $\mathfrak{A} \cong_{\kappa}^{\alpha} \mathfrak{B}$, if there is a sequence $\{I_\beta : \beta \leq \alpha\}$ of non-empty sets of isomorphisms from submodels of $\mathfrak{A}$ onto submodels of $\mathfrak{B}$ such that

$$I_0 \supseteq I_1 \supseteq \dots \supseteq I_\alpha,$$

and also satisfying the following back-and-forth conditions:

for any $\beta + 1 \leq \alpha$, $f \in I_{\beta+1}$ and $C \subseteq A$ with $|C| < \kappa$ (or, $D \subseteq B$ with $|D| < \kappa$), there is some $g \in I_\beta$ such that $f \subseteq g$ and $C \subseteq \mathrm{dom}(g)$ (or, $D \subseteq \mathrm{ran}(g)$).

It is convenient to consider the empty model as a submodel of any model for a language with no individual constant symbols. Therefore the empty function is an isomorphism between submodels of any two models for such a language.

$I_\alpha$ only needs to contain one function, namely an isomorphism of the submodel of $\mathfrak{A}$ generated by the constants onto the corresponding submodel of $\mathfrak{B}$, which is the empty function if the language contains no constants. So, $\mathfrak{A} \cong_\kappa^0 \mathfrak{B}$ iff the submodel of $\mathfrak{A}$ generated by the constants is isomorphic to the corresponding submodel of $\mathfrak{B}$. Also, $\mathfrak{A} \cong_\kappa^1 \mathfrak{B}$ iff $\mathfrak{A}$ and $\mathfrak{B}$ have, up to isomorphism, the same submodels generated by fewer than $\kappa$ elements.

Clearly, $\mathfrak{A} \cong_\kappa \mathfrak{B}$ implies $\mathfrak{A} \cong_\kappa^\alpha \mathfrak{B}$ for all $\alpha$, since given $I: \mathfrak{A} \cong_\kappa \mathfrak{B}$ we can let $I_\beta = I$ for all $\beta \leq \alpha$ in the definition. The converse is true but harder, and is more easily derived from the results of the next chapter.

Just as for $\cong_2$ and $\cong_\kappa$, one can show that $\cong_\kappa^\alpha$ is an equivalence relation, and also that it is preserved by direct products and reduced products (see also [2], [3], [8]).

Note that, although $\cong_2$ is the same as $\cong_\omega$, $\cong_2^\alpha$ is not usually the same as $\cong_\omega^\alpha$.

The relations $\cong_2^\alpha$ were introduced by Karp [12] as a refinement of $\cong_2$, and are close to the Ehrenfeucht-Fraissé games used to show elementary equivalence with respect to finitary first-order logic.

As an example consider $\langle Q,+\rangle$ and $\langle Re,+\rangle$, where $Q$ is the set of rationals and $Re$ is the set of reals. Then it is not true that $\langle Q,+\rangle \cong_2 \langle Re,+\rangle$; this follows, for example, from the Corollary to 1.4 since $\langle Q,+\rangle$ has only countably many automorphisms. But, as a non-trivial exercise, the reader is invited to show that $\langle Q,+\rangle \cong_2^\omega \langle Re,+\rangle$. This is stronger than saying that $\langle Q,+\rangle$ and $\langle Re,+\rangle$ are elementarily equivalent (by Theorem II.2.1 (b)).

Karp's original application of these refined relations was the theorem below on well-orderings. We showed in section one that if $\mathfrak{A} = \langle A,<^{\mathfrak{A}}\rangle$ is a well-ordering, then any model partially isomorphic to $\mathfrak{A}$ is actually isomorphic to $\mathfrak{A}$. The next theorem shows, on the other hand, that for every $\alpha$ there is some well-ordered $\mathfrak{A}$

which is 2-partially isomorphic up to $\alpha$ to some model which is not well-ordered. This will yield the non-characterizability of well-order in $L_{\infty\omega}$ (Theorem II.3.1).

Let $\alpha$ be any non-zero ordinal and let $\mathfrak{A} = \langle A,<\rangle$ be any linear ordering with a least element 0. We define the ordered product $\alpha \otimes \mathfrak{A}$ to be the model $\langle \alpha \times A,<\rangle$ where $\langle \xi_1,a_1\rangle < \langle \xi_2,a_2\rangle$ iff $a_1 < a_2$ or $a_1 = a_2$ and $\xi_1 < \xi_2$. Informally, $\alpha \otimes \mathfrak{A}$ is the linear ordering consisting of $\alpha$ repeated $\mathfrak{A}$ times.

In what follows, $\omega^\beta$ denotes ordinal exponentiation. In particular, if $\kappa$ is an uncountable cardinal then $\omega^\beta < \kappa$ for all $\beta < \kappa$.

3.1 THEOREM (Karp [12]). *If* $\omega^\beta < \alpha$ *for all* $\beta < \alpha$, *then* $\langle\alpha,<\rangle \cong_2^\alpha \alpha \otimes \mathfrak{A}$.

PROOF: If $\beta < \alpha$ then an $\omega^\beta$ interval of $\alpha$ is any interval of the form

$$\{\gamma : \omega^\beta\cdot\xi \leq \gamma < \omega^\beta\cdot(\xi+1)\}.$$

An $\omega^\beta$-interval of $\alpha \otimes \mathfrak{A}$ is any interval of the form

$$\{\langle\gamma,a\rangle : \langle\omega^\beta\cdot\xi,a\rangle \leq \langle\gamma,a\rangle < \langle\omega^\beta\cdot(\xi+1),a\rangle\}.$$

For any $\beta \leq \alpha$ we define $I_\beta$ to be the collection of all functions f which are $h|s$ for some finite s and some order-preserving map h of a union of finitely many $\omega^\beta$ intervals of $\alpha$ onto a union of $\omega^\beta$ intervals of $\alpha \otimes \mathfrak{A}$ with $h(0) = \langle 0,0\rangle$. Then each $I_\beta$ is a non-empty set of isomorphisms of submodels of $\langle\alpha,<\rangle$ onto submodels of $\alpha \otimes \mathfrak{A}$, and $\beta_1 \leq \beta_2 \leq \alpha$ implies $I_{\beta_2} \subseteq I_{\beta_1}$.

We must verify the back-and-forth conditions of the definition. So, let $\beta + 1 \leq \alpha$, let $f \in I_{\beta+1}$, and pick $\langle\xi,a\rangle \in \alpha \times A$. We must find $g \in I_\beta$ with $f \subseteq g$ and $\langle\xi,a\rangle$ in the range of g. First of all, let $f = h|s$ where $h$ is an isomorphism of $\omega^{\beta+1}$ intervals. If $<\xi,a>$ is in the range of $h$, say $f(\gamma) = <\xi,a>$, then we can set

$$g = f \cup \{<\gamma,<\xi,a>>\}$$

and then $g \in I_{\beta+1}$, so in particular $g \in I_\beta$.

If $<\xi,a>$ is not in the range of $h$, then it is not in the same $\omega^{\beta+1}$ interval as any element of the range of $f$. So let $\gamma_1$ be the largest element of the domain of $f$ with $f(\gamma_1) < <\xi,a>$, and let $\gamma_2$ be the smallest element of the domain of $f$ with $<\xi,a> < f(\gamma_2)$. Then $\gamma_1$ and $\gamma_2$ are in different $\omega^{\beta+1}$ intervals, so there are infinitely many $\omega^\beta$ intervals strictly between $\gamma_1$ and $\gamma_2$. Choose any one of these to map isomorphically onto the $\omega^\beta$ interval containing $<\xi,a>$; this takes some $\gamma$ to $<\xi,a>$. Then the function

$$g = f \cup \{<\gamma,<\xi,a>>\}$$

is in $I_\beta$.

The other back-and-forth condition can be similarly verified.

⊣

As particular examples, we have:

(a) $<\omega_1,<> \cong_2^{\omega_1} <\omega_1+\omega_1,<> \cong_2^{\omega_1} <\kappa,<>$ for all uncountable cardinals $\kappa$.

(b) $<\kappa,<> \cong_2^{\kappa} <\kappa+\kappa\cdot\omega^*,<>$ for all uncountable cardinals $\kappa$.

In particular, (b) shows that for every uncountable cardinal $\kappa$, the well-ordering $<\kappa,<>$ is 2-partially isomorphic up to $\kappa$ to some non-well-order of cardinality $\kappa$.

A more detailed study of well-orderings and the relations $\cong_2^\alpha$ is contained in [18]. For other related material see [3], [8], and [9].

# CHAPTER II

## INFINITARY FORMULAS

1. The Infinitary Logics $L_{\kappa\lambda}$. In this section we finally define the classes of infinitary formulas to be studied in the rest of this book. Our starting point is a (finitary) language $L$ with equality and an arbitrary collection (perhaps empty) of predicate symbols, function symbols and individual constant symbols. The infinitary formulas are built up from $L$ using:

(*i*) a proper class of individual variables, say $v_\alpha$ for every ordinal $\alpha$;

(*ii*) the logical connectives $\neg$ (not), $\wedge$ (conjunction) and $\vee$ (disjunction);

(*iii*) the quantifiers $\forall$ (for all) and $\exists$ (there exists).

The *terms* and *atomic formulas* are defined as for finitary logic (although we have a proper class of each due to having a proper class of individual variables).

The class $L_{\infty\infty}$ of infinitary formulas is defined as the least class containing the atomic formulas and such that:

(1) if $\varphi$ is a formula then so is $\neg\varphi$;

(2) if $\Phi$ is a set of formulas, then $\wedge\Phi$ and $\vee\Phi$ are formulas;

(3) if $\varphi$ is a formula and $W$ is a set of variables then $\forall W\varphi$ and $\exists W\varphi$ are formulas.

We pick out certain subclasses of $L_{\infty\infty}$ by cardinality restrictions on the sets allowed in (2) and (3).

If $\lambda$ is any cardinal greater than or equal to 2, then $L_{\infty\lambda}$ is the class obtained by restricting the set $W$ of variables in (3) to have cardinality less than $\lambda$.

If $\kappa$ is an infinite cardinal and $\kappa \geq \lambda$, then $L_{\kappa\lambda}$ is the class of formulas of $L_{\infty\lambda}$ obtained by restricting the set $\Phi$ of formulas in (2) to have cardinality less than $\kappa$.

The first extensive treatment of these logics was in Karp's book [11]. A recent account is in Dickmann's forthcoming book [7]. The very important case of $L_{\omega_1\omega}$ receives detailed treatment in Keisler's book [14].

The difference between $L_{\kappa\omega}$ and $L_{\kappa 2}$ is very slight: in $L_{\kappa 2}$ we must quantify one variable at a time, while in $L_{\kappa\omega}$ we may quantify finitely many variables at once. These two logics, therefore, will turn out to be semantically equivalent, the only differences being those that arise when we count the number of quantifiers in a formula.

The reader should be warned that in most other accounts, the logic called $L_{\kappa\omega}$ is really what we call $L_{\kappa 2}$ here.

Notice that $L_{\omega\omega}$, as well as $L_{\omega 2}$, is essentially the usual finitary logic, which allows only finite conjunctions and disjunctions and finite quantification.

Since the formulas are inductively defined, we have available the techniques of proof by induction and definition by induction over formulas of $L_{\infty\infty}$.

Occurrences of variables in formulas are characterized as *free* or *bound* by the obvious extension of the definition for finitary logic. By a *sentence* we mean a formula with no free variables. If $\varphi$ is a formula and $\vec{x}$ is a sequence of variables containing all the free variables of $\varphi$, then we will also write $\varphi$ as $\varphi(\vec{x})$. If $\vec{t}$ is some sequence of terms of the same length as $\vec{x}$, then $\varphi(\vec{t})$

is the formula obtained by replacing all free occurrences of $x_\xi$ in $\varphi$ by $t_\xi$, for every $\xi$, after first changing the names of bound variables in $\varphi$ so as to avoid quantifying any variable occurring in $\vec{t}$.

The definition of *satisfaction* of formulas in a model is a straightforward generalization of the usual finitary definition. Let $\mathfrak{A}$ be a model for L. By an *assignment* in $\mathfrak{A}$ we mean a function $\underset{\sim}{a}$ whose domain is some set of variables and whose range is a subset of A. We inductively define $\mathfrak{A} \models \varphi[\underset{\sim}{a}]$ for all formulas $\varphi$ and assignments $\underset{\sim}{a}$ defined on all the free variables of $\varphi$ as follows:

*(i)* $\varphi$ atomic -- just as finitary logic;

*(ii)* $\mathfrak{A} \models \neg\varphi[\underset{\sim}{a}]$ iff not $(\mathfrak{A} \models \varphi[\underset{\sim}{a}])$;

*(iii)* $\mathfrak{A} \models \bigwedge\Phi[\underset{\sim}{a}]$ iff $\mathfrak{A} \models \varphi[\underset{\sim}{a}]$ for all $\varphi \in \Phi$;

*(iv)* $\mathfrak{A} \models \bigvee\Phi[\underset{\sim}{a}]$ iff $\mathfrak{A} \models \varphi[\underset{\sim}{a}]$ for some $\varphi \in \Phi$;

*(v)* $\mathfrak{A} \models \forall W\varphi[\underset{\sim}{a}]$ iff $\mathfrak{A} \models \varphi[\underset{\sim}{b}]$ for every assignment $\underset{\sim}{b}$ defined on all free variables of $\varphi$ such that $\underset{\sim}{b}$ agrees with $\underset{\sim}{a}$ on all free variables of $\forall W\varphi$;

*(vi)* $\mathfrak{A} \models \exists W\varphi[\underset{\sim}{a}]$ iff $\mathfrak{A} \models \varphi[\underset{\sim}{b}]$ for some assignment $\underset{\sim}{b}$ defined on all free variables of $\varphi$ such that $\underset{\sim}{b}$ agrees with $\underset{\sim}{a}$ on all free variables of $\exists W\varphi$.

If $\varphi$ is $\varphi(\vec{x})$ for a sequence $\vec{x} = \langle x_\xi : \xi < \alpha \rangle$ of variables and $\vec{a}$ is the sequence $\langle \underset{\sim}{a}(x_\xi) : \xi < \alpha \rangle$ then in place of $\mathfrak{A} \models \varphi[\underset{\sim}{a}]$ we normally write $\mathfrak{A} \models \varphi[\vec{a}]$.

If $\mathfrak{A} \models \varphi[\vec{a}]$ for all sequences $\vec{a}$ from A then we write simply $\mathfrak{A} \models \varphi$. $\varphi$ is *logically valid*, written $\models \varphi$, if $\mathfrak{A} \models \varphi$ for all $\mathfrak{A}$. Two formulas are *logically equivalent* if they are satisfied by exactly the same sequences in every model.

The reader can easily verify that the following pairs of formulas are logically equivalent:

$$\bigwedge\Phi, \quad \neg\bigvee\{\neg\varphi : \varphi\in\Phi\};$$

$$\forall W\varphi, \quad \neg\exists W\neg\varphi;$$

$$\exists(V\cup W)\varphi, \quad \exists V\exists W\varphi;$$

$$\forall(V\cup W)\varphi, \quad \forall V\forall W\varphi.$$

These last two equivalences imply that a quantifier on finitely many variables can be equivalently replaced by several quantifiers each on one variable. Therefore *every formula of* $L_{\kappa\omega}$ *is logically equivalent to a formula of* $L_{\kappa 2}$. For this reason we usually refer to $L_{\kappa\omega}$ instead of $L_{\kappa 2}$ except when we need to distinguish formulas by their quantifier-rank (defined below).

$\mathfrak{A} \equiv_{\infty\lambda} \mathfrak{B}$, and $\mathfrak{A} \equiv_{\kappa\lambda} \mathfrak{B}$, mean that $\mathfrak{A} \models \varphi$ iff $\mathfrak{B} \models \varphi$ for all sentences $\varphi$ of $L_{\infty\lambda}$, and $L_{\kappa\lambda}$, respectively.

The set $\mathrm{sub}(\varphi)$ of *subformulas* of $\varphi$ is defined inductively as follows:

$$\mathrm{sub}(\varphi) = \{\varphi\} \quad \text{if } \varphi \text{ is atomic;}$$

$$\mathrm{sub}(\neg\varphi) = \{\neg\varphi\} \cup \mathrm{sub}(\varphi);$$

$$\mathrm{sub}(\bigwedge\Phi) = \{\bigwedge\Phi\} \cup \bigcup\{\mathrm{sub}(\varphi) : \varphi\in\Phi\};$$

$$\mathrm{sub}(\bigvee\Phi) = \{\bigvee\Phi\} \cup \bigcup\{\mathrm{sub}(\varphi) : \varphi\in\Phi\};$$

$$\mathrm{sub}(\forall W\varphi) = \{\forall W\varphi\} \cup \mathrm{sub}(\varphi);$$

$$\mathrm{sub}(\exists W\varphi) = \{\exists W\varphi\} \cup \mathrm{sub}(\varphi).$$

It is easily proved that if $\kappa$ is regular then

$$L_{\kappa\lambda} = \{\varphi\in L_{\infty\lambda} : |\mathrm{sub}(\varphi)| < \kappa\}$$

If $\kappa$ is singular then every formula of $L_{\kappa^+\lambda}$ is equivalent to a formula of $L_{\kappa\lambda}$. The point is just that conjunctions and disjunctions of sets of cardinality $\kappa$ can be replaced by conjunctions and disjunctions of sets of cardinality less than $\kappa$ -- if $\Phi \subseteq L_{\kappa\lambda}$ and $|\Phi| = \kappa$ then $\Phi = \bigcup_{i\in I} \Phi_i$ where $|I| < \kappa$ and $|\Phi_i| < \kappa$ for all $i \in I$, and hence $\wedge\Phi$ is equivalent to $\wedge\{\wedge\Phi_i : i\in I\}$, which is a formula of $L_{\kappa\lambda}$. Because of this we will ignore $L_{\kappa\lambda}$ for singular $\kappa$.

It is a straightforward induction to show that if $\lambda' = \max(\lambda,\omega)$ and if $\varphi$ is a formula of $L_{\infty\lambda}$ with fewer than $\lambda'$ free variables and $\psi \in \mathrm{sub}(\varphi)$, then $\psi$ also has fewer than $\lambda'$ free variables. In particular subformulas of sentences of $L_{\infty\lambda}$ all have fewer than $\lambda'$ free variables. Because of this fact we add the following restriction to the definition of $L_{\infty\lambda}$:

> *every formula of* $L_{\infty\lambda}$ *contains fewer than* $\max(\lambda,\omega)$ *free variables.*

$L_{\kappa\lambda}$ is a proper class, but one can ask how many inequivalent sentences it contains. If $\kappa$ is regular, then any sentence $\varphi$ of $L_{\kappa\lambda}$ has fewer than $\kappa$ subformulas, each with fewer than $\kappa$ free variables, so $\varphi$ contains fewer than $\kappa$ variables in all. So every sentence of $L_{\kappa\lambda}$ is equivalent to one using just the variables $v_\alpha$ for $\alpha < \kappa$. If $L$ has at most $\kappa$ symbols it is clear that there can be at most $\kappa^{\underset{\smile}{\kappa}}$ such formulas. Therefore $L_{\kappa\lambda}$ contains at most $\kappa^{\underset{\smile}{\kappa}}$ inequivalent sentences. Since $\kappa^{+\underset{\smile}{\kappa}^{+}} = 2^\kappa$ we see in particular that $L_{\kappa^+\lambda}$ contains at most $2^\kappa$ inequivalent sentences. We will see that this bound is the best possible, even when $\lambda = \omega$.

If we are given some indexing, say $\{\varphi_i : i\in I\}$, of the set $\Phi$ of formulas, then we shall often write $\bigwedge_{i\in I} \varphi_i$ instead of $\wedge\Phi$ and $\bigvee_{i\in I} \varphi_i$ instead of $\vee\Phi$. We will also write $\varphi\wedge\psi$ and $\varphi\vee\psi$ in place of $\wedge\{\varphi,\psi\}$ and $\vee\{\varphi,\psi\}$ respectively. We define $\longrightarrow$ and $\longleftrightarrow$ just as for finitary logic. Similarly, if the set $W$ of variables is enumerated, say as $\{w_\xi : \xi < \alpha\}$, then we may write $(\exists w_\xi)_{\xi<\alpha}\varphi$, or

$(\exists w_0 w_1 \ldots)\varphi$, in place of $\exists W\varphi$, and analogously for $\forall W\varphi$.

We now give some examples of what can be expressed in various $L_{\kappa\lambda}$'s.

For the first examples we let $L$ have an individual constant symbol $\overline{1}$ and a binary function symbol $+$.

(1) There is a sentence of $L_{\omega_1\omega}$ characterizing the model $\langle P,+,1\rangle$ up to isomorphism, where $P$ is the set of positive integers. Let $\overline{n}$ be the term $\overline{1}+\ldots+\overline{1}$, with $n$ occurrences of $\overline{1}$. Then the sentence can be taken to be

$$\forall x(\bigvee_{n\in P} x=\overline{n}) \wedge \bigwedge_{m,n\in P} [\overline{m}+\overline{n} = (\overline{m+n})] \wedge \bigwedge_{m\neq n} \overline{m}\neq\overline{n}.$$

(2) There is a sentence of $L_{\omega_1\omega}$ characterizing the model $\langle Q^+,+,1\rangle$ up to isomorphism, where $Q^+$ is the set of positive rationals. Let $x = (\overline{m/n})$ be an abbreviation for $\overline{m} = x+\ldots+x$, with $n$ occurrences of $x$. The required sentence then states that every element is $\overline{m/n}$ for some $m,n\in P$, that for all $m,n\in P$ some element is $\overline{m/n}$, and that addition and equality on such elements $\overline{m/n}$ behaves as in $Q^+$. The details of translating this are left to the reader.

(3) There is a sentence of $L_{\kappa\omega}$, where $\kappa = (2^{\omega})^+$, and also a sentence of $L_{\omega_1\omega_1}$, characterizing the model $\langle R^+,+,1\rangle$ up to isomorphism, where $R^+$ is the set of positive reals. Here we leave all the work to the reader.

For the next examples we take $L$ to have just a binary relation $<$ as non-logical symbol.

(4) For any non-zero ordinal $\alpha$ such that $\alpha<\kappa$ there is a sentence of $L_{\kappa\omega}$ characterizing $\langle\alpha,<\rangle$ up to isomorphism. First we define formulas $\psi_\nu(x)$ by induction as follows:

$\psi_0(x)$ is $\neg\exists y\,(y < x)$

$\psi_\nu(x)$ is $\forall y[y < x \longleftrightarrow \bigvee_{\xi<\nu} \psi_\xi(y)]$ for $\nu \neq 0$.

Let $\theta$ be the finitary sentence saying that $<$ is a linear order. Then in models of $\theta$, $\psi_\nu(x)$ says that the predecessors of $x$ are linearly ordered like $\nu$. Finally let $\sigma_\alpha$ be

$$\forall x(\bigvee_{\nu<\alpha} \psi_\nu(x)) \wedge \bigwedge_{\nu<\alpha} \exists x \psi_\nu(x) \wedge \theta.$$

Then $\sigma_\alpha$ is easily shown to be the desired sentence.

Note that it follows from this that $\langle A,<\rangle$ is well-ordered iff $\langle A,<\rangle \models \sigma_\alpha$ for some $\alpha$. In particular, if $\langle A,<\rangle$ is well-ordered and $\langle A,<\rangle \equiv_{\infty\omega} \langle B,<'\rangle$ then they are isomorphic.

Let I, J be distinct subsets of $\kappa$ each of cardinality less than $\kappa$. Then $\bigvee_{\alpha\in I} \sigma_\alpha$ and $\bigvee_{\alpha\in J} \sigma_\alpha$ are inequivalent sentences of $L_{\kappa\omega}$. Since there are $\kappa^{\underline{\kappa}}$ such subsets we therefore see that $L_{\kappa\omega}$ has $\kappa^{\underline{\kappa}}$ inequivalent sentences.

(5) $\langle Q,<\rangle \equiv_{\infty\omega} \langle Re,<\rangle$, where Q is the set of rationals and Re is the set of reals. This will follow from the results of the next section.

(6) There is a sentence $\sigma$ of $L_{\omega_1\omega_1}$ such that $\langle A,<\rangle \models \sigma$ iff $<$ well-orders A. Let $\sigma$ say that $<$ is a linear order and $\neg(\exists x_n)_{n<\omega} \bigwedge_{k<\omega} (x_{k+1} < x_k)$. We will see in section three of this chapter that such a sentence cannot be found in $L_{\infty\omega}$.

For any language L note the following.

(7) There is a sentence $\sigma$ of $L_{\omega_1\omega_1}$ such that $\mathfrak{A} \models \sigma$ iff $|A| \leq \omega$. As $\sigma$ we can take

$$(\exists x_n)_{n<\omega} \forall y \bigvee_{k<\omega} (y = x_k).$$

It will follow from the results of the next section that there is no such sentence in $L_{\infty\omega}$.

(8) If $\mathfrak{A}$ is any model of power $\lambda$ then there is a sentence of $L_{\infty\lambda^+}$ determining $\mathfrak{A}$ up to isomorphism. This easy fact is left to the reader.

(9) If $\mathfrak{A}_i \equiv_{\infty\omega} \mathfrak{B}_i$ for $i = 1,2$, then $\mathfrak{A}_1 \times \mathfrak{A}_2 \equiv_{\infty\omega} \mathfrak{B}_1 \times \mathfrak{B}_2$. This will follow using results from the next section. However, there are $\mathfrak{A}_i \equiv_{\omega_1\omega} \mathfrak{B}_i$ for $i = 1,2$ such that $\mathfrak{A}_1 \times \mathfrak{A}_2 \not\equiv_{\omega_1\omega} \mathfrak{B}_1 \times \mathfrak{B}_2$ (Malitz [20]).

One useful measure of the complexity of a formula is its *quantifier-rank*, defined inductively as follows:

$$\begin{aligned} qr(\varphi) &= 0 \quad \text{if } \varphi \text{ is atomic} \\ qr(\neg\varphi) &= qr(\varphi) \\ qr(\forall W\varphi) &= qr(\exists W\varphi) = qr(\varphi) + 1 \\ qr(\wedge\Phi) &= qr(\vee\Phi) = \sup(\{qr(\varphi) : \varphi\in\Phi\}). \end{aligned}$$

Notice that $\exists x_1x_2\varphi$ and $\exists x_1\exists x_2\varphi$, though logically equivalent, have different quantifier-ranks. In general, although a formula of $L_{\kappa\omega}$ has a natural translation into $L_{\kappa 2}$, this translation may have higher quantifier-rank than the original formula.

We write $L^{\alpha}_{\infty\lambda}$ for the class of formulas of $L_{\infty\lambda}$ with quantifier-rank less than or equal to $\alpha$. We write $\mathfrak{A} \equiv^{\alpha}_{\infty\lambda} \mathfrak{B}$ to indicate that $\mathfrak{A}$ and $\mathfrak{B}$ satisfy the same sentences of $L^{\alpha}_{\infty\lambda}$. Our remarks above show that in general $\mathfrak{A} \equiv^{\alpha}_{\infty 2} \mathfrak{B}$ does not imply $\mathfrak{A} \equiv^{\alpha}_{\infty\omega} \mathfrak{B}$.

It is easy to see that if $\kappa$ is a regular cardinal then $L_{\kappa\lambda} \subseteq L^{\kappa}_{\infty\lambda}$, but the reverse inclusion fails. From some points of view this classification by quantifier rank is more natural than that by cardinality of conjunctions and disjunctions.

The examples we have given show that a sentence of $L_{\kappa\omega}$, for example, with an infinite model need not have models of other infinite cardinalities. However, we do have the following quite satisfactory version of the downward Löwenheim-Skolem theorem.

1.1 THEOREM (Hanf). *Assume* $|L| \leq \kappa$. *Let* $\sigma$ *be a sentence of* $L_{\kappa^+\lambda}$ *and let* $\mu$ *be a cardinal such that* $\kappa \leq \mu = \mu^{\underset{\smile}{\lambda}}$. *Then for any* $\mathfrak{A}$ *and any* $A_0 \subseteq A$ *with* $|A_0| \leq \mu \leq |A|$ *there is some* $\mathfrak{B} \subseteq \mathfrak{A}$ *with* $A_0 \subseteq B$, $|B| = \mu$, *and such that for every subformula* $\varphi(\vec{x})$ *of* $\sigma$ *and every* $\vec{b}$ *from* $B$

$$\mathfrak{B} \models \varphi[\vec{b}] \quad \textit{iff} \quad \mathfrak{A} \models \varphi[\vec{b}].$$

*In particular,* $\mathfrak{B} \models \sigma$ *iff* $\mathfrak{A} \models \sigma$.

PROOF: This is similar to the proof of the downward Löwenheim-Skolem theorem in finitary logic. That is, we define Skolem functions for the quantifiers in $\sigma$ and close $A_0$ under them. The details are trickier because of the more complicated structure of infinitary sentences and because of cardinality considerations.

First of all, we can assume $\lambda \geq \omega$. Secondly, we may assume $\sigma$ contains no universal quantifiers, by replacing $\forall\vec{w}$ by $\neg\exists\vec{w}\neg$ everywhere in $\sigma$. Next let $S'$ be the set of all subformulas of such a $\sigma$. Then $|S| \leq \kappa$ and every formula in $S$ has fewer than $\lambda$ free variables. Let $\varphi$ in $S$ begin with a quantifier, so $\varphi$ is $\exists\vec{y}\psi(\vec{x},\vec{y})$ where $\vec{x}$ is a sequence of $\alpha < \lambda$ variables and $\vec{y}$ is a sequence of $\beta < \lambda$ variables. We define a sequence $\langle f_\xi^\varphi : \xi < \beta\rangle$ of $\beta$ functions each of $\alpha$ arguments so that for any $\alpha$-termed sequence $\vec{a}$ from $A$,

$$\mathfrak{A} \models \varphi[\vec{a}] \quad \text{iff} \quad \mathfrak{A} \models \psi[\vec{a},\vec{b}]$$

where $b_\xi = f_\xi^\varphi(\vec{a})$ for all $\xi < \beta$. This is, of course, easy with the axiom of choice. Now let $F$ be the set of all such functions $f_\xi^\varphi$ with $\varphi \in S$. Then $|F| \leq \kappa$ since each formula of $S$ contributes at most $\kappa$ functions to $F$.

If $\mathfrak{B} \subseteq \mathfrak{A}$ is such that $B$ is closed under all the functions in $F$, then an easy induction on $S$ shows that for all $\varphi(\vec{x})$ in $S$ and all $\vec{b}$ from $B$

$$\mathfrak{B} \models \varphi[\vec{b}] \quad \text{iff} \quad \mathfrak{A} \models \varphi[\vec{b}].$$

Given $A_0$ and $\mu$ as in the statement of the theorem we may assume that $|A_0| = \mu$. Let $B$ be the closure of $A_0$ under the functions in $F$ and the interpretations of the function symbols of $L$. Then $|B| = \mu$ since $\mu = \mu^{\underset{\smile}{\lambda}}$ and we are closing under at most $\mu$ functions each with fewer than $\lambda$ arguments. Therefore the submodel $\mathfrak{B}$ of $\mathfrak{A}$ with universe $B$ is the required model.

The following consequence is a simplification of the theorem which contains enough information for many purposes.

COROLLARY. *A sentence of* $L_{\kappa^+\omega}$ *which has a model has a model of cardinality at most* $\kappa$. *A sentence of* $L_{\kappa^+\lambda^+}$ *which has a model has a model of cardinality at most* $\kappa^\lambda$.

Note also that if a sentence of $L_{\kappa^+\omega}$ has a model of power $\lambda$ where $\kappa < \lambda$, then it has models of all powers between $\kappa$ and $\lambda$. This property is not true for $L_{\infty\lambda}$ with $\omega < \lambda$. For example, the reader can find a sentence $\sigma$ of $L_{\omega_1\omega_1}$ such that $\sigma$ has a model of power $\kappa$ if and only if $\kappa^\omega = \kappa$.

The situation with an upward Löwenheim-Skolem theorem is not as easy. Since there are only $2^\kappa$ inequivalent sentences of $L_{\kappa^+\lambda}$ we know there is some cardinal $\mu$ such that a sentence of $L_{\kappa^+\lambda}$ with a model of cardinality at least $\mu$ has models of arbitrarily large powers (Hanf's theorem). Determining $\mu$, even in simple cases, goes beyond this paper. We mention just that for $L_{\omega_1\omega}$, $\mu$ is known to be $\beth_{\omega_1}$. For further information see [5], [7].

There are, finally, the following notions of elementary submodels with respect to infinitary formulas.

Definition. $\mathfrak{A} \prec_{\infty\lambda} \mathfrak{B}$, $\mathfrak{A}$ is an $L_{\infty\lambda}$-*elementary submodel of* $\mathfrak{B}$, if $\mathfrak{A} \subseteq \mathfrak{B}$ and for every formula $\varphi(\vec{x})$ of $L_{\infty\lambda}$ and for any sequence $\vec{a}$ from A,

$$\mathfrak{A} \models \varphi[\vec{a}] \quad \text{iff} \quad \mathfrak{B} \models \varphi[\vec{a}].$$

$\mathfrak{A} \prec_{\kappa\lambda} \mathfrak{B}$ and $\mathfrak{A} \prec^{\alpha}_{\infty\lambda} \mathfrak{B}$ are defined by restricting $\varphi$ to be in $L_{\kappa\lambda}$ or $L^{\alpha}_{\infty\lambda}$, respectively.

The following version of the Tarski-Vaught union of elementary chains theorem is easily established and left to the reader. The reason we need $\lambda \leq cf(\mu)$ is so that any sequence of fewer than $\lambda$ elements from the union already belongs to some model of the chain.

1.2 THEOREM. *Let* $\{\mathfrak{A}_\xi\}_{\xi<\mu}$ *be a chain of models such that* $\mathfrak{A}_{\xi_1} \prec_{\infty\lambda} \mathfrak{A}_{\xi_2}$ *whenever* $\xi_1 < \xi_2 < \mu$, *and assume that* $\lambda \leq cf(\mu)$. *Then*

$$\mathfrak{A}_{\xi_0} \prec_{\infty\lambda} \bigcup_{\xi<\mu} \mathfrak{A}_\xi \quad \textit{for every} \quad \xi_0 < \mu.$$

*The same holds for* $\prec_{\kappa\lambda}$ *and* $\prec^{\alpha}_{\infty\lambda}$.

Since every well-ordering is determined up to isomorphism by a sentence of $L_{\infty\omega}$, we know that no well-ordering has a proper $L_{\infty\omega}$-elementary submodel or extension. The downward Löwenheim-Skolem theorem does yield the following, however.

1.3 THEOREM. *Let* $\mathfrak{A}$ *be a model for a language* L *with at most* $\kappa$ *symbols. Then there is some* $\mathfrak{B} \prec_{\kappa^+\kappa^+} \mathfrak{A}$ *with* $|B| \leq 2^\kappa$.

PROOF: $L_{\kappa^+\kappa^+}$ has at most $2^\kappa$ inequivalent sentences. Let $\sigma$ be the conjunction of all of them. Then $\sigma$ is a sentence of $L_{(2^\kappa)^+,\kappa^+}$. Assume that $|A| \geq 2^\kappa$. Then we can set $\mu = 2^\kappa$ in Theorem 1.1, since $(2^\kappa)^{\kappa^+} = 2^\kappa$, which then yields $\mathfrak{B} \subseteq \mathfrak{A}$ with $|B| \leq 2^\kappa$. Since every formula of $L_{\kappa^+\kappa^+}$ is equivalent to a subformula of $\sigma$, the conclusion of Theorem 1.1 states precisely that $\mathfrak{B} \prec_{\kappa^+\kappa^+} \mathfrak{A}$, as desired. $\dashv$

Examples are easily given indicating that this result is best possible.

2. <u>The Connections with Partial Isomorphisms</u>. In this section we give the results which link the concepts of $\kappa$-partial isomorphisms from Chapter I with the infinitary logics introduced in the preceding section. The fundamental result here is Theorem 2.1 below, due primarily to Carol Karp. In [12] Karp explicitly introduced the relations $\cong_2$ and $\cong_2^\alpha$ and proved that case of Theorem 2.1 below. Various people, certainly including Karp, soon realized that there were analogous back-and-forth characterizations for $L_{\infty\kappa}$ and $L_{\infty\kappa}^\alpha$ for arbitrary $\kappa$. This is implicit in Chang [5], for example. These characterizations first appear explicitly in Benda [2], and independently in Calais [3]. Both of these papers apply the characterizations to obtain results stating that various algebraic operations preserve certain infinitary equivalences.

2.1 THEOREM (Karp). *Let* $\kappa \geq 2$ *be given. Then:*

(a) $\mathfrak{A} \cong_\kappa \mathfrak{B}$ *iff* $\mathfrak{A} \equiv_{\infty\kappa} \mathfrak{B}$

(b) $\mathfrak{A} \cong_\kappa^\alpha \mathfrak{B}$ *iff* $\mathfrak{A} \equiv_{\infty\kappa}^\alpha \mathfrak{B}$, *for any ordinal* $\alpha$.

PROOF: Let $\kappa' = \max(\kappa,\omega)$. In part (a) we could assume that $\kappa \geq \omega$, but it is necessary to consider finite $\kappa$ for part (b).

(a) First assume that

$$I: \mathfrak{A} \cong_\kappa \mathfrak{B} .$$

We show by induction on formulas $\varphi(\vec{x})$ of $L_{\infty\kappa}$, where $\vec{x}$ is a sequence of $\mu < \kappa'$ variables, that for any sequence $\langle a_\xi \rangle_{\xi<\mu}$ from $A$ and for any $f \in I$ with $a_\xi \in \mathrm{dom}(f)$ for all $\xi < \mu$

$$\mathfrak{A} \models \varphi[\vec{a}] \quad \text{iff} \quad \mathfrak{B} \models \varphi[\langle f(a_\xi) \rangle_{\xi<\mu}].$$

If $\varphi$ is atomic this is clear, since functions in $I$ are isomorphisms and hence preserve atomic formulas.

The inductive steps corresponding to the connectives are trivial, so we just do the quantifiers. Say $\varphi$ is $\exists \vec{y} \psi(\vec{x},\vec{y})$, where $\vec{y}$ is a sequence of $\alpha < \kappa$ variables and $\psi$ has the desired property. Let $\vec{a}$ be a $\mu$-termed sequence from $A$ and let $f \in I$ be defined at every $a_\xi$, $\xi < \mu$. Assume that $\mathfrak{A} \models \varphi[\vec{a}]$. Then $\mathfrak{A} \models \psi[\vec{a},\vec{c}]$, where $\vec{c}$ is an $\alpha$-termed sequence from $A$. By the $\kappa$-back-and-forth property, since $\alpha < \kappa$, there is some $g \in I$ such that $g \supseteq f$ and $c_\xi \in \mathrm{dom}(g)$ for every $\xi < \alpha$. Then, by our inductive hypothesis,

$$\mathfrak{B} \models \psi[\langle g(a_\xi) \rangle_{\xi<\mu}, \langle g(c_\xi)_{\xi<\alpha} \rangle].$$

That is, $\mathfrak{B} \models \varphi[\langle f(a_\xi) \rangle_{\xi<\mu}]$ as desired, since $f(a_\xi) = g(a_\xi)$ for all $\xi < \mu$. We similarly show that if $\mathfrak{B} \models \varphi[\langle f(a_\xi) \rangle_{\xi<\mu}]$ then $\mathfrak{A} \models \varphi[\vec{a}]$.

Since the universal quantifier is definable from negation and the existential quantifier, this completes the induction and shows that for every sentence $\sigma$ of $L_{\infty\kappa}$, $\mathfrak{A} \models \sigma$ iff $\mathfrak{B} \models \sigma$.

Now for the other direction, the main step is to show the following:

(#) If $\mathfrak{A}^*$ and $\mathfrak{B}^*$ are models for some language $L^*$,

$\mathfrak{A}^* \equiv_{\infty\kappa} \mathfrak{B}^*$ and $\vec{a}$ is a sequence of $< \kappa$ elements of $A^*$, then there is a sequence $\vec{b}$ from $B^*$ such that $(\mathfrak{A}^*,\vec{a}) \equiv_{\infty\kappa} (\mathfrak{B}^*,\vec{b})$.

Say that $\vec{a}$ is a sequence of length $\mu$. Let $S$ be the set of all sequences $\vec{b}$ of length $\mu$ from $B^*$ such that $(\mathfrak{A}^*,\vec{a}) \not\equiv_{\infty\kappa} (\mathfrak{B}^*,\vec{b})$. For every $\vec{b} \in S$ let $\varphi_{\vec{b}}(\vec{x})$ be a formula of $L^*_{\infty\kappa}$ such that $\mathfrak{A}^* \models \varphi_{\vec{b}}[\vec{a}]$ but $\mathfrak{B}^* \models \neg\varphi_{\vec{b}}[\vec{b}]$. Let $\psi$ be $\bigwedge\{\varphi_{\vec{b}} : \vec{b} \in S\}$. Then $\mathfrak{A}^* \models \psi[\vec{a}]$, hence $\mathfrak{A}^* \models \exists\vec{x}\psi$, and so $\mathfrak{B}^* \models \exists\vec{x}\psi$ since this is a sentence of $L^*_{\infty\kappa}$. Let $\vec{b}$ be such that $\mathfrak{B}^* \models \psi[\vec{b}]$. Then $\vec{b}$ cannot be in $S$ since every sequence in $S$ falsifies $\psi$. Therefore by the definition of $S$ we must have $(\mathfrak{A}^*,\vec{a}) \equiv_{\infty\kappa} (\mathfrak{B}^*,\vec{b})$ as desired.

Now assume that $\mathfrak{A} \equiv_{\infty\kappa} \mathfrak{B}$. We will define a set $I$ such that $I: \mathfrak{A} \cong_{\kappa} \mathfrak{B}$. To simplify the presentation assume that $L$ has no function or individual constant symbols and so any subset of $A$ is the universe of a submodel of $\mathfrak{A}$. $I$, then, is to consist of all functions $f$ such that $\mathrm{dom}(f)$ is a subset of $A$ of cardinality less than $\kappa'$, $\mathrm{ran}(f) \subseteq B$, and if $\mathrm{dom}(f) = \{a_\xi : \xi < \mu\}$ and $b_\xi = f(a_\xi)$ for each $\xi < \mu$, then $(\mathfrak{A},\vec{a}) \equiv_{\infty\kappa} (\mathfrak{B},\vec{b})$. Then $I$ is a non-empty set of isomorphisms and (#), applied to $\mathfrak{A}^* = (\mathfrak{A},\vec{a})$ and $\mathfrak{B}^* = (\mathfrak{B},\vec{b})$ and conversely, tells us that $I$ has the $\kappa$-back-and-forth property. And so $I: \mathfrak{A} \cong_{\kappa} \mathfrak{B}$ as desired.

(b). This is in outline very much like (a). For the first direction we show inductively that functions in $I_\nu$ preserve formulas in $L^\nu_{\infty\kappa}$ for all $\nu \leq \alpha$. For the reverse direction we prove the variant of (#) where the hypothesis is $\mathfrak{A}^* \equiv^{\nu+1}_{\infty\kappa} \mathfrak{B}^*$ and the conclusion is $(\mathfrak{A}^*,\vec{a}) \equiv^{\nu}_{\infty\kappa} (\mathfrak{B}^*,\vec{b})$. Further details can be left to the reader. ⊣

In some later arguments depending on this theorem what we actually use is (#), which can be loosely expressed as saying that

$\equiv_{\infty\kappa}$ defines a $\kappa$-back-and-forth relation between elements of models.

Let us note some immediate consequences of this theorem. First of all, since we know that $\mathfrak{A} \equiv_{\infty\kappa} \mathfrak{B}$ if and only if $\mathfrak{A} \equiv^{\alpha}_{\infty\kappa} \mathfrak{B}$ for all $\alpha$, we have the following, which was remarked but not proved in Section I.3.

COROLLARY. *$\mathfrak{A} \cong_{\kappa} \mathfrak{B}$ iff $\mathfrak{A} \cong^{\alpha}_{\kappa} \mathfrak{B}$ for every* $\alpha$.

Secondly, using also the corollary to Theorem I.2.1, we obtain the next result.

COROLLARY. *If $|A| = |B| = \kappa$ and $cf(\kappa) = \omega$, then $\mathfrak{A} \equiv_{\infty\kappa} \mathfrak{B}$ iff $\mathfrak{A} \cong \mathfrak{B}$.*

Looking carefully at the proof of Theorem 2.1(a) one can see that $L_{\infty\kappa}$-elementary equivalence of $\mathfrak{A}$ and $\mathfrak{B}$ is not needed to conclude that they are $\kappa$-partially isomorphic, but that $L_{\mu^+\kappa}$-elementary equivalence suffices, where $\mu$ is some cardinal depending only on $\kappa$ and $|B|$ -- in particular $\mu$ is independent of $\mathfrak{A}$ and the size of L. The resulting theorem is the following. We can assume without loss of generality that $\kappa \geq \omega$ since we are not classifying formulas by quantifier-rank.

2.2 THEOREM. *Given infinite $\mathfrak{B}$ and $\kappa$, let $\mu = |B|^{\underset{\smile}{\kappa}}$. Then for any $\mathfrak{A}$,*

$$\mathfrak{A} \cong_{\kappa} \mathfrak{B} \quad \textit{iff} \quad \mathfrak{A} \equiv_{\mu^+\kappa} \mathfrak{B}.$$

PROOF: All we need to do is establish the following sharper version of (#):

(##) Let $\mathfrak{A}^*$ and $\mathfrak{B}^*$ be models for a language $L^*$ and assume $\mathfrak{A}^* \equiv_{\mu^+\kappa} \mathfrak{B}^*$ where $\mu = |B^*|^{\underset{\smile}{\kappa}}$. Then for any sequence $\vec{a}$ of $< \kappa$ elements of $A^*$ there is a sequence $\vec{b}$ from $B^*$ such that $(\mathfrak{A}^*,\vec{a}) \equiv_{\mu+\kappa} (\mathfrak{B}^*,\vec{b})$, and similarly given $\vec{b}$ we can find such an $\vec{a}$.

First of all, given $\vec{a}$ of length $\gamma < \kappa$ we can proceed rather as in the proof of (#); that is, we define $S$ to be the set of all sequences $\vec{b}$ of length $\gamma$ from $B^*$ such that $(\mathfrak{A}^*,\vec{a}) \not\equiv_{\mu+\kappa} (\mathfrak{B}^*,\vec{b})$. We choose for every $\vec{b}$ in $S$ a formula $\varphi_{\vec{b}}(\vec{x})$ of $L^*_{\mu+\kappa}$ such that $\mathfrak{A}^* \models \varphi_{\vec{b}}[\vec{a}]$ but $\mathfrak{B}^* \models \neg\varphi_{\vec{b}}[\vec{b}]$, and let $\psi$ be $\wedge\{\varphi_{\vec{b}}: \vec{b}\in S\}$. Then $\psi$ is also a formula of $L_{\mu+\kappa}$ since $|S| \leq |B^*|^{|\gamma|} \leq \mu$. Therefore $\mathfrak{B}^* \models \exists\vec{x}\psi$, since this is a sentence of $L^*_{\mu+\kappa}$ true on $\mathfrak{A}^*$, and any $\vec{b}$ satisfying $\psi$ will do the job.

Given $\vec{b}$ of length $\gamma < \kappa$ we must proceed differently since it may be that $|A^*|^{|\gamma|} > \mu$. So, let $S$ be the set of all sequences $\vec{c}$ of length $\gamma$ from $B^*$ such that $(\mathfrak{B}^*,\vec{b}) \not\equiv_{\mu+\kappa} (\mathfrak{B}^*,\vec{c})$. For each $\vec{c}$ in $S$ let $\varphi_{\vec{c}}(\vec{x})$ be a formula of $L^*_{\mu+\kappa}$ satisfied by $\vec{b}$ but not $\vec{c}$, and let $\psi$ be $\wedge\{\varphi_{\vec{c}}: \vec{c}\in S\}$. Then $\psi$ is a formula of $L^*_{\mu+\kappa}$ determining $\vec{b}$ up to $L^*_{\mu+\kappa}$-elementary equivalence in $\mathfrak{B}^*$; that is $\mathfrak{B}^* \models \exists\vec{x}\psi$, and for every formula $\theta(\vec{x})$ of $L^*_{\mu+\kappa}$ satisfied by $\vec{b}$ we have

$$\mathfrak{B}^* \models \forall\vec{x}(\psi \longrightarrow \theta).$$

All these sentences of $L^*_{\mu+\kappa}$ are then true in $\mathfrak{A}^*$, hence there is some $\vec{a}$ satisfying $\psi$ in $\mathfrak{A}^*$. Any such $\vec{a}$ must also satisfy every $L^*_{\mu+\kappa}$ formula satisfied by $\vec{b}$, and so $(\mathfrak{A}^*,\vec{a}) \equiv_{\mu+\kappa} (\mathfrak{B}^*,\vec{b})$ as desired.

Having established (##), the theorem follows by an argument like that used for 2.1(a). ⊣

Notice, of course, that if $\kappa = \omega$ then $\mu = |B|$ in the statement of Theorem 2.2. We thus have the following improvement of the second corollary to 2.1 for $\kappa = \omega$.

COROLLARY. *If* $|A| = |B| = \omega$ *and* $\mathfrak{A} \equiv_{\omega_1\omega} \mathfrak{B}$ *then* $\mathfrak{A} \cong \mathfrak{B}$.

Another immediate consequence of 2.2 is also worth noting here.

COROLLARY. *Given* $\mathfrak{B}$ *there is a sentence* $\sigma$ *of* $L_{\infty\kappa}$ *such that for every* $\mathfrak{A}$

$$\mathfrak{A} \models \sigma \quad \textit{iff} \quad \mathfrak{A} \equiv_{\infty\kappa} \mathfrak{B}.$$

PROOF: Let $\sigma$ be the conjunction of all the sentences of $L_{\mu^+\kappa}$ true on $\mathfrak{B}$. ⊣

In fact there is a more striking improvement in 2.1 than that given by 2.2, provided the language L is not too large. Namely a single sentence of $L_{\mu^+\kappa}$ can be found characterizing $\mathfrak{B}$ up to $L_{\infty\kappa}$-elementary equivalence. This is a generalization of Scott's theorem, given as a corollary below, that any countable model for a countable language is characterized up to isomorphism among countable models by a sentence of $L_{\omega_1\omega}$.

2.3 THEOREM. *Let* $\mathfrak{B}$ *be an infinite model for a language with at most* $|B|$ *symbols, and let* $\mu = |B|^{\underset{\smile}{\kappa}}$, *where* $\kappa \geq \omega$. *Then there is a sentence* $\sigma$ *of* $L_{\mu^+\kappa}$ *such that for every* $\mathfrak{A}$

$$\mathfrak{A} \models \sigma \quad \textit{iff} \quad \mathfrak{A} \equiv_{\infty\kappa} \mathfrak{B}.$$

PROOF: We give two proofs of this theorem. The first is fairly easy and starts from the proof we have given of Theorem 2.2. Given $\mathfrak{B}$ and $\mu$ as in the statement of this theorem, let $\vec{b}$ be any sequence of elements of B of length $\alpha < \kappa$. The proof of (##) in 2.2 yields a formula $\psi_{\vec{b}}(\vec{x})$ of $L_{\mu^+\kappa}$ such that for every $\alpha$-termed sequence $\vec{c}$ from B

$$\mathfrak{B} \models \psi_{\vec{b}}[\vec{c}] \quad \text{iff} \quad (\mathfrak{B},\vec{b}) \equiv_{\infty\kappa} (\mathfrak{B},\vec{c}).$$

By the $\kappa$-back-and-forth property of $L_{\infty\kappa}$-elementary equivalence, we know the following are all true on $\mathfrak{B}$:

$$(i)\quad \forall\vec{x}[\psi_{\vec{b}}(\vec{x}) \longrightarrow \wedge\{\exists\vec{y}\psi_{\vec{b},\vec{c}}(\vec{x},\vec{y}) : \vec{c}\in S\}],$$

$$(ii)\quad \forall\vec{x}[\psi_{\vec{b}}(\vec{x}) \longrightarrow \forall\vec{y}\vee\{\psi_{\vec{b},\vec{c}}(\vec{x},\vec{y}) : \vec{c}\in S\}],$$

where $S$ is the set of all $<\kappa$-termed sequences from $B$ and $\vec{b} \in S$, and

$$(iii)\quad \forall\vec{x}[\psi_{\vec{b}}(\vec{x}) \longrightarrow \wedge\Gamma_{\vec{b}}]$$

where $\Gamma_{\vec{b}}$ is the set of all atomic and negated atomic formulas satisfied by $\vec{b}$ in $\mathfrak{B}$.

Each of these sentences is a sentence of $L_{\mu^+\kappa}$, since $|S| = |B|^{\underline{\kappa}} = \mu$ and $L$ contains at most $\mu$ symbols (this last is needed just for (*iii*)). There are also only $\mu$ sentences of the form (*i*), (*ii*) or (*iii*), once again since $|S| = \mu$. Hence the conjunction of all the sentences (*i*), (*ii*) and (*iii*) is a sentence $\sigma$ of $L_{\mu^+\kappa}$. Now, given $\mathfrak{A} \models \sigma$ it is easy to show by induction on formulas $\varphi(\vec{x})$ of $L_{\infty\kappa}$ that for any $\vec{a}$ from $A$ and $\vec{b}$ from B:

$$\text{if } \mathfrak{A} \models \psi_{\vec{b}}[\vec{a}] \text{ then } \mathfrak{B} \models \varphi[\vec{b}] \text{ iff } \mathfrak{A} \models \varphi[\vec{a}].$$

For atomic $\varphi$ this is given by the sentences (*iii*), the connectives are trivial, and (*i*) and (*ii*) enable us to deal with the quantifiers. In particular, setting $\vec{b} = \phi$, we see that $\mathfrak{A} \equiv_{\infty\kappa} \mathfrak{B}$. Therefore this $\sigma$ is the sentence needed for the theorem.

The second proof, due to Chang [5], does not use the proofs of the preceeding theorems and is of more theoretical interest (for example, see the Technical Remark on page 20 of [1]). As in the previous proof, we let $S$ be the set of all sequences of elements of $B$ of length less than $\kappa$, and for $\vec{b} \in S$ we let $\Gamma_{\vec{b}}$ be the set of all atomic and negated atomic formulas satisfied by $\vec{b}$ in $\mathfrak{B}$. For every $\vec{b} \in S$ we define formulas $\varphi^{\alpha}_{\vec{b}}(\vec{x})$ by induction on $\alpha$ as follows:

$$\varphi^{0}_{\vec{b}} \quad \text{is} \quad \wedge\Gamma_{\vec{b}};$$

$$\varphi^{\alpha+1}_{\vec{b}} \quad \text{is} \quad \varphi^{\alpha}_{\vec{b}} \wedge \wedge\{\exists y \varphi^{\alpha}_{\vec{b},\vec{c}}(\vec{x},\vec{y}) : \vec{c} \in S\} \wedge \forall \vec{y}\ \vee\{\varphi^{\alpha}_{\vec{b},\vec{c}}(\vec{x},\vec{y}) : \vec{c} \in S\};$$

$$\varphi^{\alpha}_{\vec{b}} \quad \text{is} \quad \bigwedge_{\beta<\alpha} \varphi^{\beta}_{\vec{b}} \quad \text{if} \quad \alpha \quad \text{is a limit ordinal.}$$

An easy induction, using the facts that $|S| = \mu$ and the language has at most $\mu$ symbols shows that for every $\alpha < \mu^{+}$, $\varphi^{\alpha}_{\vec{b}}$ is a formula of $L_{\mu^{+}\kappa}$ of quantifier-rank $\alpha$ and $\mathfrak{B} \models \varphi^{\alpha}_{\vec{b}}[\vec{b}]$. Also notice that if $\beta \leq \alpha$ then $\models \forall \vec{x}(\varphi^{\alpha}_{\vec{b}} \longrightarrow \varphi^{\beta}_{\vec{b}})$. Since $|S| = \mu$ there is some ordinal $\gamma < \mu^{+}$ such that for every $\vec{b} \in S$ and every $\alpha$, if $\gamma \leq \alpha < \mu^{+}$ then

$$\mathfrak{B} \models \forall \vec{x}(\varphi^{\gamma}_{\vec{b}} \longleftrightarrow \varphi^{\alpha}_{\vec{b}}).$$

We define $\sigma$ to be

$$\varphi^{\gamma}_{\phi} \wedge \wedge\{\forall \vec{x}(\varphi^{\gamma}_{\vec{b}} \longrightarrow \varphi^{\gamma+1}_{\vec{b}}) : \vec{b} \in S\}.$$

Then $\sigma$ is a sentence of $L_{\mu^{+}\kappa}$ of quantifier-rank $\gamma+2$, and $\mathfrak{B} \models \sigma$. Now assume that $\mathfrak{A} \models \sigma$. We show by induction on formulas $\psi(\vec{x})$ of $L_{\infty\kappa}$ that for any $\vec{a}$ and $\vec{b}$ from B

$$\text{if } \mathfrak{A} \models \varphi^{\gamma}_{\vec{b}}[\vec{a}] \text{ then } \mathfrak{A} \models \psi[\vec{a}] \text{ iff } \mathfrak{B} \models \psi[\vec{b}].$$

The only difficulty in this is the case of a quantifier. So, let $\psi$ be $\exists \vec{y}\theta(\vec{x},\vec{y})$ and assume that $\mathfrak{A} \models \varphi^{\gamma}_{\vec{b}}[\vec{a}] \wedge \psi[\vec{a}]$. Then there is some $\vec{c}$ from A such that $\mathfrak{A} \models \theta[\vec{a},\vec{c}]$. Since $\mathfrak{A} \models \forall \vec{x}(\varphi^{\gamma}_{\vec{b}} \longrightarrow \varphi^{\gamma+1}_{\vec{b}})$ we actually have $\mathfrak{A} \models \varphi^{\gamma+1}_{\vec{b}}[\vec{a}]$. By the definition of $\varphi^{\gamma+1}$ there must be some $\vec{d}$ from B such that $\mathfrak{A} \models \varphi^{\gamma}_{\vec{b},\vec{d}}[\vec{a},\vec{c}]$. The inductive hypothesis on $\theta$ then implies that $\mathfrak{B} \models \theta[\vec{b},\vec{d}]$, that is, $\mathfrak{B} \models \psi[\vec{b}]$

as required. The reverse direction is similarly established. Thus, setting $\vec{b} = \emptyset$, we obtain $\mathfrak{A} \equiv_{\infty\kappa} \mathfrak{B}$. ⊣

A sentence $\sigma$ as in the statement of Theorem 2.3 is often called a *Scott sentence* for $\mathfrak{B}$. The case $\kappa = \omega$ of the following corollary in Scott's Theorem [23]; the general case is due to Chang [5].

COROLLARY. *If* $|B| = \kappa$, $cf(\kappa) = \omega$, *and* L *has at most* $\kappa$ *symbols, then there is a sentence* $\sigma$ *of* $L_{\mu^+\kappa}$, *where* $\mu = \kappa^{\underset{\smile}{\kappa}}$, *such that for every* $\mathfrak{A}$ *of cardinality at most* $\kappa$,

$$\mathfrak{A} \models \sigma \quad \textit{iff} \quad \mathfrak{A} \cong \mathfrak{B}.$$

There is also a theorem similar to the preceding one for the relation $\equiv^{\alpha}_{\infty\kappa}$.

2.4 THEOREM. *For any* $\mathfrak{B}$, *cardinal* $\kappa \geq 2$, *and ordinal* $\alpha$ *there is a sentence* $\sigma$ *of* $L^{\alpha}_{\infty\kappa}$ *such that for all* $\mathfrak{A}$

$$\mathfrak{A} \models \sigma \quad \textit{iff} \quad \mathfrak{A} \equiv^{\alpha}_{\infty\kappa} \mathfrak{B}.$$

PROOF: Let us assume $\kappa \geq \omega$; the changes necessary for finite $\kappa$ are trivial. We then proceed as in Chang's proof of 2.3 to define the formulas $\varphi^{\alpha}_{\vec{b}}$. Note that $\varphi^{\alpha}_{\vec{b}}$ has quantifier-rank $\alpha$. Given $\mathfrak{A}$ it is then easy to show that if $\mathfrak{A} \models \varphi^{\alpha}_{\vec{b}}[a]$ then $(\mathfrak{A},\vec{a}) \equiv^{\alpha}_{\infty\kappa} (\mathfrak{B},\vec{b})$. In particular, the sentence $\varphi^{\alpha}_{\emptyset}$ characterizes $\mathfrak{B}$ up to $L^{\alpha}_{\infty\kappa}$-elementary equivalence. ⊣

As an easy consequence of Theorem 2.1 we have the following characterization of $L_{\infty\kappa}$-elementary submodels.

2.5 THEOREM. *Let* $\kappa$ *be an infinite cardinal. Then* $\mathfrak{A} \prec_{\infty\kappa} \mathfrak{B}$ *iff for every sequence* $\vec{a}$ *of fewer than* $\kappa$ *elements of* A *we have*

$(\mathfrak{A},\vec{a}) \cong_\kappa (\mathfrak{B},\vec{a})$.

By Theorem 2.1 and examples from Chapter I, we know there are countable models which are $L_{\infty\omega}$-elementarily equivalent to uncountable models, and thus not isomorphic to them. The next corollary says something about the relationship they must have.

COROLLARY. *If* $\mathfrak{A} \equiv_{\infty\omega} \mathfrak{B}$ *and* $|A| = \omega$, *then* $\mathfrak{A} \cong \mathfrak{A}'$ *for some* $\mathfrak{A}' \prec_{\infty\omega} \mathfrak{B}$.

PROOF: Let $A = \{a_n : n\in\omega\}$. Since $\mathfrak{A} \cong_\omega \mathfrak{B}$, by 2.1(a), we can find $b_n \in B$, for all $n \in \omega$, such that

$$(\mathfrak{A},a_0,\ldots,a_n) \cong_\omega (\mathfrak{B},b_0,\ldots,b_n) \quad \text{for all } n \in \omega.$$

Let $\mathfrak{A}'$ be the submodel of $\mathfrak{B}$ whose universe is $\{b_n : n\in\omega\}$. Then $\mathfrak{A} \cong \mathfrak{A}'$ under the obvious mapping, and 2.5 implies that $\mathfrak{A}' \prec_{\infty\omega} \mathfrak{B}$.

Results analogous to Theorems 2.1 and 2.2 but dealing with more restricted classes of infinitary formulas, and having cofinality $\omega$ consequences concerning homomorphism and embedding rather than isomorphism, are found in [5].

3. Some Applications of Partial Isomorphisms to Infinitary Model Theory. This section contains applications of the results of the preceeding section to the model theory of infinitary logic. We emphasize $L_{\infty\omega}$ since the results are more striking there.

The simplest applications are direct translations of results on partial isomorphism to results on the corresponding infinitary logic. For example, we showed in section I.2 that $\langle Re,<\rangle$ is determined up to isomorphism by $\omega_1$-partial isomorphism, hence by Theorem 2.3, there is a sentence of $L_{\mu^+\omega_1}$, where $\mu = 2^\omega$, determining $\langle Re,<\rangle$

up to isomorphism. It is also easy, of course, to write down such a sentence directly. On the other hand, <Re,<> is not determined up to isomorphism by $L_{\infty\omega}$-elementary equivalence since it is partially isomorphic to any other dense linear order without endpoints.

More interestingly, we showed in Section I.1 that if $\mathfrak{A}_i \cong_2 \mathfrak{B}_i$, for $i = 1, 2$, then $\mathfrak{A}_1 \times \mathfrak{A}_2 \cong_2 \mathfrak{B}_1 \times \mathfrak{B}_2$, and we remarked in I.2 and I.3 that the same is true for $\cong_\kappa$ and $\cong_\kappa^\alpha$. We therefore obtain that $\equiv_{\infty\omega}$, $\equiv_{\infty\kappa}$, and $\equiv_{\infty\kappa}^\alpha$ are preserved by direct products. In fact, they are also preserved by infinite direct products, reduced products and other similar algebraic operations ([2], [3]; for a general theorem see [8]). This should be contrasted with Malitz's result that $\equiv_{\kappa\lambda}$ is preserved by direct products iff $\kappa$ is strongly inaccessible [20].

The following interesting negative theorem uses the results on ranked partial isomorphisms. Stronger versions have been obtained by other methods (e.g., in [17]).

3.1 THEOREM (Karp [12], López-Escobar [17]). *Let* L *be a language with a binary relation* <. *There is no sentence* $\sigma$ *of* $L_{\infty\omega}$ *such that for all* $\mathfrak{A}$, $\mathfrak{A} \models \sigma$ *iff* $<^{\mathfrak{A}}$ *well-orders* A.

PROOF: If there were such a sentence it would belong to $L_{\infty 2}^\alpha$ for some $\alpha$, and so be preserved by $\cong_2^\alpha$ by Theorem 2.1(b). But by Theorem I.3.1 there is for any $\alpha$ some well-ordering which is partially isomorphic up to $\alpha$ to some non-well-order.

We are also now in a position to turn the results on automorphisms of countable models from Section I.1 into definability theorems. The main fact needed is the following, which is contained in the proof of Theorem 2.2 and is here explicitly stated as a lemma.

LEMMA. *Let* $\mathfrak{A}$ *be countable and let* $a \in A$. *Then there is a formula* $\varphi(x)$ *of* $L_{\omega_1\omega}$ *such that for any* $c \in A$, $\mathfrak{A} \models \varphi[c]$ *iff* $(\mathfrak{A},a) \cong (\mathfrak{A},c)$.

If the element $a$ is fixed by every automorphism of $\mathfrak{A}$, then $a$ must be the only element of $A$ satisfying the formula $\varphi$. Consequently we have the following, which is due to Scott in the case of a countable language. The definability condition simply says that every element of $A$ is $L_{\omega_1\omega}$-definable.

3.2 THEOREM (Scott [23]). *Let* $\mathfrak{A}$ *be countable. Then* $\mathfrak{A}$ *has no proper automorphisms iff there are formulas* $\varphi_n(x)$ *of* $L_{\omega_1\omega}$, $n \in \omega$, *such that*

$$\mathfrak{A} \models \forall x \bigvee_{n\in\omega} \varphi_n(x) \quad \wedge \bigwedge_{n\in\omega} \exists! x \varphi_n(x).$$

Combining Theorems 3.2 and I.1.4 we immediately obtain the next theorem.

3.3 THEOREM (Kueker [15]). *Let* $\mathfrak{A}$ *be countable. Then the following are equivalent:*

(*i*) $\mathfrak{A}$ *has* $< 2^{\omega}$ *automorphisms*

(*ii*) $\mathfrak{A}$ *has* $\leq \omega$ *automorphisms*

(*iii*) *There are formulas* $\varphi_n(z_1,\ldots,z_k,x)$ *of* $L_{\omega_1\omega}$, $n \in \omega$, *such that*

$$\mathfrak{A} \models \exists z_1 \ldots z_k [\forall x \bigvee_{n\in\omega} \varphi_n \wedge \bigwedge_{n\in\omega} \exists! x \varphi_n].$$

The corollary to I.1.4 directly translates to the following using 2.2.

COROLLARY (Kueker [15]). *If* $\mathfrak{A}$ *is countable and is* $L_{\omega_1\omega}$-

*elementarily equivalent to some uncountable model, then* $\mathfrak{A}$ *has* $2^{\omega}$ *automorphisms.*

There is a similar pair of results on defining subsets of countable models. The next theorem corresponds to 3.2 and is also due to Scott, for countable languages. ($\underset{\sim}{P}$ is a unary predicate not in L).

3.4 THEOREM (Scott [23]). *Let* $\mathfrak{A}$ *be a countable model for* L *and let* $P \subseteq A$. *The following are then equivalent:*

(*i*) $(\mathfrak{A},P) \cong (\mathfrak{A},Q)$ *implies that* $P = Q$, *for all* $Q \subseteq A$.

(*ii*) *There is a formula* $\psi(x)$ *of* $L_{\omega_1\omega}$ *such that*

$$(\mathfrak{A},P) \models \forall x[\underset{\sim}{P}(x) \longleftrightarrow \psi(x)].$$

PROOF: (*ii*) $\Longrightarrow$ (*i*) is obvious. Assuming (*i*) we show (*ii*). For every $a \in P$ let $\varphi_a(x)$ be the formula of $L_{\omega_1\omega}$ given by the lemma. Let $c \in A$ and assume that $\mathfrak{A} \models \varphi_a[c]$. Then there is an automorphism h of $\mathfrak{A}$ such that $h(a) = c$. Let $Q = \{h(b) : b \in P\}$. Then h is an isomorphism of $(\mathfrak{A},P)$ onto $(\mathfrak{A},Q)$, so $P = Q$. Therefore $c \in P$. So, P is the set of all elements satisfying $\varphi_a(x)$ for some $a \in P$, and hence (*ii*) holds for $\psi$ defined as $\bigvee_{a \in P} \varphi_a$.

Combining Theorems I.1.5 and 3.4 we similarly obtain the definability version of I.1.5. We leave to the reader the precise formulation of this result, and also the versions of these results which hold for models of cardinality cofinal with $\omega$ (see [15]).

The remaining results generally concern $L_{\infty\omega}$-elementary equivalence, especially of models of different cardinalities.

We first prove a simple characterization of the countable

models which are $L_{\infty\omega}$-elementarily equivalent to uncountable models. Makkai [19] has given an interesting explicit axiomatization of the countable models which are $L_{\infty\omega}$-elementarily equivalent to uncountable models and applied it to linear orderings.

3.5 THEOREM. *Let* $\mathfrak{A}$ *be countable. Then* $\mathfrak{A}$ *is* $L_{\infty\omega}$*-elementarily equivalent to an uncountable model iff* $\mathfrak{A}$ *has a proper* $L_{\omega_1\omega}$*-elementary extension.*

PROOF: If $\mathfrak{A} \equiv_{\infty\omega} \mathfrak{B}$ then $\mathfrak{A}$ is isomorphic to an $L_{\infty\omega}$-elementary submodel of $\mathfrak{B}$, by the corollary to 2.5. In particular, if $\mathfrak{B}$ is uncountable then $\mathfrak{A}$ has an uncountable $L_{\infty\omega}$-elementary extension. For the other direction, let $\mathfrak{A} \prec_{\omega_1\omega} \mathfrak{A}_1$ where $\mathfrak{A} \neq \mathfrak{A}_1$. If $\mathfrak{A}_1$ is uncountable we are done (by Theorem 2.2), so assume $\mathfrak{A}_1$ is countable. Then $\mathfrak{A} \cong \mathfrak{A}_1$. So $\mathfrak{A}_1$ has a proper countable $L_{\omega_1\omega}$-elementary extension $\mathfrak{A}_2$, and this has a proper countable $L_{\omega_1\omega}$-elementary extension $\mathfrak{A}_3$, etc. $\mathfrak{A}_\omega = \bigcup_{n\in\omega} \mathfrak{A}_n$ is an $L_{\omega_1\omega}$-elementary extension of each $\mathfrak{A}_n$ by Theorem 1.2, so we can continue for $\omega_1$ steps. The resulting model $\mathfrak{B} = \bigcup_{\xi\in\omega_1} \mathfrak{A}_\xi$ is an uncountable $L_{\omega_1\omega}$-elementary extension of $\mathfrak{A}$, and so is $L_{\infty\omega}$-elementarily equivalent to $\mathfrak{A}$.

The next two results require the Scott-type improvements given by Theorem 2.3, since we will use the downward Löwenheim-Skolem theorem.

The first makes precise the sense in which a proof that any two countable models of a theory are isomorphic is also a proof that any two models of the theory are partially isomorphic.

3.6 THEOREM. *Let* $\sigma$ *be a sentence of* $L_{\omega_1\omega}$ *where* L *is countable. Then any two countable models of* $\sigma$ *are isomorphic iff any two models of* $\sigma$ *are partially isomorphic.*

PROOF: We only need to prove the implication from left to right. Let $\mathfrak{A}$ be the countable model of $\sigma$, and let $\sigma^*$ of $L_{\omega_1\omega}$ be the Scott sentence of $\mathfrak{A}$, that is a sentence determining $\mathfrak{A}$ up to $L_{\infty\omega}$-elementary equivalence (which exists by 2.3). If there were a model of $\sigma$ which is not partially isomorphic to $\mathfrak{A}$ it would be a model of $\sigma\wedge\neg\sigma^*$. By Theorem 1.2, $\sigma\wedge\neg\sigma^*$ would have a countable model -- but this would be a countable model of $\sigma$ not isomorphic to $\mathfrak{A}$, a contradiction.

The second gives a sufficient condition for all models of a sentence of $L_{\infty\omega}$ to be $L_{\infty\omega}$-elementarily equivalent to "small" models, that is models of some bounded cardinality. The proof is a straightforward generalization of the preceding proof.

3.7 THEOREM. *Let* $\sigma$ *be a sentence of* $L_{\kappa^+\omega}$, *where* L *has at most* $\kappa$ *symbols. Assume that* $\sigma$ *has at most* $\kappa$ *non* $L_{\infty\omega}$-*elementarily equivalent models of power* $\kappa$. *Then every model of* $\sigma$ *is* $L_{\infty\omega}$-*elementarily equivalent to a model of power at most* $\kappa$.

PROOF: Let $\{\mathfrak{A}_\xi\}_{\xi<\kappa}$ list the non $L_{\infty\omega}$-elementarily equivalent models of $\sigma$ of power $\kappa$, and let $\sigma_\xi$ of $L_{\kappa^+\omega}$ be the Scott sentence of $\mathfrak{A}_\xi$ (by 2.3). Let $\sigma^*$ be $\bigvee_{\xi<\kappa}\sigma_\xi$. Then $\sigma^*$ is a sentence of $L_{\kappa^+\omega}$, and every model of $\sigma$ of power $\kappa$ is also a model of $\sigma^*$. So by the downward Löwenheim-Skolem theorem, every model of $\sigma$ of power greater than $\kappa$ is also a model of $\sigma^*$, and therefore is $L_{\infty\omega}$-elementarily equivalent to some $\mathfrak{A}_\xi$, $\xi < \kappa$. ⊣

The final result gives an interesting characterization of the models which are $L_{\infty\omega}$-elementarily equivalent to models of power at most $\kappa$.

3.8 THEOREM. *For any* $\mathfrak{A}$ *the following are equivalent.*

(*i*) $\mathfrak{A} \equiv_{\infty\omega} \mathfrak{B}$ *for some* $\mathfrak{B}$ *with* $|B| \leq \kappa$.

(*ii*) $\mathfrak{B} \prec_{\infty\omega} \mathfrak{A}$ *for some* $\mathfrak{B}$ *with* $|B| \leq \kappa$.

(*iii*) $\mathfrak{A}$ *has at most* $\kappa$ *n-tuples which are* $L_{\infty\omega}$*-inequivalent, for each* $n \in \omega$.

PROOF: (*ii*) ⟹ (*i*) ⟹ (*iii*) are obvious. Assuming (*iii*) we will show (*ii*) as follows. By the back-and-forth properties of $L_{\infty\omega}$-elementary equivalence there are functions $f_n : A^{2n+1} \longrightarrow A$, $n \in \omega$, such that

if $(\mathfrak{A}, a_0, \ldots, a_{n-1}) \equiv_{\infty\omega} (\mathfrak{A}, b_0, \ldots, b_{n-1})$

then $(\mathfrak{A}, a_0, \ldots, a_{n-1}, a_n) \equiv_{\infty\omega} (\mathfrak{A}, b_0, \ldots, b_{n-1}, f_n(a_0, \ldots, a_n, b_0, \ldots, b_{n-1}))$.

By (*iii*), there is some $A_0 \subseteq A$ with $|A_0| \leq \kappa$ such that for any $a_0, \ldots, a_{n-1} \in A$ there are $b_0, \ldots, b_{n-1} \in A_0$ with

$$(\mathfrak{A}, a_0, \ldots, a_{n-1}) \equiv_{\infty\omega} (\mathfrak{A}, b_0, \ldots, b_{n-1}).$$

Let $B$ be the closure of $A_0$ under $\{f_n : n \in \omega\}$. Then $|B| \leq \kappa$ and $B$ is the universe of some $\mathfrak{B} \subseteq \mathfrak{A}$. A completely straightforward induction on formulas $\varphi(x_1, \ldots, x_n)$ of $L_{\infty\omega}$ shows that

$$\mathfrak{A} \models \varphi[b_1, \ldots, b_n] \quad \text{iff} \quad \mathfrak{B} \models \varphi[b_1, \ldots, b_n]$$

for all $b_1, \ldots, b_n \in B$. Therefore $\mathfrak{B} \prec_{\infty\omega} \mathfrak{A}$ and so (*ii*) holds. ⊣

This brief survey has omitted many interesting and important applications of back-and-forth methods. One is Scott's original use of Scott's Theorem to prove a result on invariant Borel sets [24]. We have already referred to Fefferman's general theorem on operations preserving infinitary equivalences [8]. There are also several applications to abelian groups; see, for example, [1]. Keisler

[13] has used back-and-forth techniques to prove equivalence in even stronger infinitary logics, allowing linearly ordered quantifiers. Finally, there is Morley's theorem that a sentence $\sigma$ of $L_{\omega_1\omega}$ either has at most $\omega_1$ non-isomorphic countable models or has exactly $2^{\omega}$ non-isomorphic countable models [21]. The proof of this theorem uses a construction like Chang's proof of Theorem 2.3 to analyze the $L_{\omega_1\omega}$-types consistent with $\sigma$. The reader should consult these papers for a broader understanding of the possible uses of these techniques.

## REFERENCES


1. K.J. Barwise, *Back and forth through infinitary logic*, in: Studies in Model Theory, MAA Studies vol. 8, 1973, 5-34.

2. M. Benda, *Reduced products and nonstandard logics*, Journal of Symbolic Logic 34(1969), 424-436.

3. J.-P. Calais, *Partial isomorphisms and infinitary languages*, Zeitschrift für Math. Logik 18(1972), 435-456.

4. G. Cantor, Contributions to the Founding of the Theory of Transfinite Numbers, Dover Publ. Co., New York.

5. C.C. Chang, *Some remarks on the model theory of infinitary languages*, in: The Syntax and Semantics of Infinitary Languages, Springer-Verlag, Berlin, 1968, 36-63.

6. C.C. Chang and H.J. Keisler, Model Theory, North-Holland Publ. Co., Amsterdam, 1974.

7. M. Dickmann, Large Infinitary Languages, North-Holland Publ. Co., Amsterdam, to appear.

8. S. Fefferman, *Infinitary properties, local functors, and systems of ordinal functions*, in: Conference in Mathematical Logic - London '70, Springer-Verlag, Berlin, 1972, 63-97.

9. R. Fraïssé, *Isomorphisme local et équivalence associés à un ordinal; utilité en calcul des formules infinies à quanteurs finis*, in: Proceedings of the Tarski Symposium, American Mathematical Society, Providence, 1974, 241-254.

10. F. Hausdorff, Grundzüge der Mengenlehre, Chelsea Publ. Co., New York.

11. C. Karp, Languages with Expressions of Infinite Length, North-Holland Publ. Co., Amsterdam, 1964.

12. C. Karp, *Finite quantifier equivalence*, in: The Theory of Models, North-Holland Publ. Co., Amsterdam, 1965, 407-412.

13. H.J. Keisler, *Formulas with linearly ordered quantifiers*, in: The Syntax and Semantics of Infinitary Languages, Springer-Verlag, Berlin, 1968, 96-130.

14. H.J. Keisler, Model Theory for Infinitary Logic, North-Holland Publ. Co., Amsterdam, 1971.

15. D.W. Kueker, *Definability, automorphisms, and infinitary languages*, in: The Syntax and Semantics of Infinitary Languages, Springer-Verlag, Berlin, 1968, 152-165.

16. D.W. Kueker, *Löwenheim-Skolem and interpolation theorems in infinitary languages*, Bulletin of the American Mathematical Society 78 (1972), 211-215.

17. E.G.K. López-Escobar, *On defining well-orderings*, Fundamenta Mathematicae 59(1966), 13-21.

18. E.G.K. López-Escobar, *Well-orderings and finite quantifiers*, Journal of the Mathematical Society of Japan 20(1968), 477-489.

19. M. Makkai, *Structures elementarily equivalent relative to infinitary languages to models of higher power*, Acta Mathematica Acad. Sci. Hungar. 21(1970), 283-295.

20. J. Malitz, *Infinitary analogs of theorems from first order model theory*, Journal of Symbolic Logic 36(1971), 216-228.

21. M. Morley, *The number of countable models*, Journal of Symbolic Logic 35(1970), 14-18.

22. G.E. Reyes, *Local definability theory*, Annals of Math. Logic 1 (1969), 101-138.

23. D. Scott, *Logic with denumerably long formulas*, in: The Theory of Models, North-Holland Publ. Co., Amsterdam, 1965, 329-341.

24. D. Scott, *Invariant Borel sets*, Fundamenta Mathematicae 56(1964), 117-128.

25. T. Skolem, *Logisch-kombinatorische Untersuchungen über die Erfüllbarkeit und Beweisbarkeit mathematischen Sätze nebst einem Theoreme über dichte Mengen*, in: Selected Works in Logic by T. Skolem, Universitetsforlaget, Oslo, 1970, 103-136.

26. W. Tait, *Equivalence in* $L_{\infty\lambda}$ *and isomorphism*, to appear.

PART B

# CONSISTENCY PROPERTIES FOR FINITE QUANTIFIER LANGUAGES

BY

JUDY GREEN

# CONTENTS
## PART B

# INTRODUCTION†

This article presents an expansion of Smullyan's "unifying principle" for quantification theory [25] to languages which allow uncountable conjunctions and disjunctions. As Smullyan's abstract consistency properties for finitary first-order languages and Makkai's generalization of these to $L_{\omega_1\omega}$ [22] can be used to prove interpolation, completeness and compactness theorems in a model-theoretic fashion, so can the $\kappa$ consistency properties defined here. Although Smullyan's consistency property is a $\kappa$ consistency property, Makkai's is not. However it is included in this presentation as a model for our generalizations. Since Keisler's Model Theory for Infinitary Logic [18] served as an inspiration for these generalizations we use it as a guide.

The uncountable languages considered here fall into two types, the original infinitary languages classified by cardinality and those developed by Barwise [1] using definability criteria. The first type gives us completeness theorems and the generalizations of the Craig, Lyndon and Malitz interpolation theorems for languages $L_{\kappa\omega}$, $\kappa$ regular, and $L_{\kappa^+\omega}$, $\kappa$ of cofinality $\omega$, while the second type gives us an uncountable generalization of the Barwise compactness theorem, the cofinality $\omega$ compactness theorem of Barwise and Karp [2], [16]. Thus the unifying principle Smullyan found for finitary logic will be seen to be a unifying principle among finite quantifier languages.

Consistency properties can also be defined for other types of languages to obtain similar results, e.g. another generalization of

---

†Most of the techniques for uncountable languages appeared in the author's doctoral dissertation written under the direction of Carol Karp. The author was partially supported by NSF Grant GP 11263 at the University of Maryland and, during preparation of this article, by the Rutgers University Research Council.

the Barwise compactness theorem [9] which is proved using, though not stated for, a modified version of the indexed languages of Karp, and interpolation and compactness theorems for chain models for infinite quantifier languages ([6], [17] and the article by Cunningham in this volume [7]).

# CHAPTER I

## LANGUAGES CLASSIFIED BY CARDINALITY

In this chapter we consider the languages $L_{\kappa\lambda}$ defined in the previous article in this volume. We use standard terminology and notation, including the following.

When we consider a language $L_{\kappa\omega}$ we assume there are less than $\kappa$ symbols of the language if $\kappa$ is a successor cardinal and at most $\kappa$ symbols if $\kappa$ is a limit cardinal.

A cardinal $\kappa$ is a *strong limit cardinal* if for all $\lambda < \kappa$, $2^{\lambda} < \kappa$. For any cardinal $\kappa$, $2^{*\kappa} = \bigcup\{(2^{\lambda})^{+} \mid \lambda < \kappa\}$ so that $\kappa$ is a strong limit cardinal iff $\kappa = 2^{*\kappa}$. A cardinal is *strongly inaccessible* if it is inaccessible and a strong limit cardinal, i.e., a regular strong limit cardinal.

If $L$ is a language including the set of constant symbols $C$ then a *basic term* of $L$ w.r.t. $C$ is a constant symbol of $L$ or a term of the form $\underline{f}(c_1,\ldots,c_n)$ where $\underline{f}$ is an $n$ place function symbol and $c_1,\ldots,c_n \in C$.

$\varphi\neg$ denotes the formula obtained by moving a negation inside $\varphi$, i.e.,

$$\varphi\neg = \neg\varphi \quad \text{if } \varphi \text{ is atomic}$$

$$(\neg\varphi)\neg = \varphi$$

$$(\bigwedge\Phi)\neg = \bigvee\{\neg\varphi \mid \varphi\in\Phi\}$$

$$(\bigvee\Phi)\neg = \bigwedge\{\neg\varphi \mid \varphi\in\Phi\}$$

$$(\forall x\varphi)\neg = \exists x\neg\varphi$$

$$(\exists x\varphi)\neg = \forall x\neg\varphi$$

$\times\{A_i \mid i\in I\}$ is the set of all functions $f$ defined on $I$ such that for all $i \in I$, $f(i) \in A_i$.

1. Model Existence Theorems. We define three types of consistency properties.

(*i*) For languages $L_{\omega_1\omega}$, called ordinary consistency properties.

(*ii*) For languages $L_{\kappa\omega}$, $\kappa$ regular, called $\kappa$ consistency properties, hence for $L_{\omega\omega}$ we have $\omega$ consistency properties.

(*iii*) For languages $L_{\kappa\omega}$, $\kappa$ singular of cofinality $\omega$, also called $\kappa$ consistency properties. This third type will allow us to prove theorems about $L_{\kappa^+\omega}$ since for $\kappa$ singular these languages have the same expressive power.

Each consistency property is a collection of sets of sentences, and is so named because being an element of one is equivalent to having a model. To show this we use a Henkin style construction of a model from constants and add to each language L a new set of constants C to form the language M. The cardinality of C depends on the language we are considering so that C is countable when we talk of ordinary consistency properties and has cardinality $\kappa$ when we talk of $\kappa$ consistency properties. In the cofinality $\omega$ case we write C as a countable union of smaller sets $C_n$.

1.1 Definitions. (*i*) Suppose S is a set of countable sets of sentences of $M_{\omega_1\omega}$. S is an *ordinary consistency property* iff for each $s \in S$ all of the following hold:

(C1) (Consistency rule) If $\varphi$ is a sentence of $M_{\omega_1\omega}$, either $\varphi \notin s$ or $\neg\varphi \notin s$.

(C2) ($\neg$ rule) If $\neg\neg\varphi \in s$, then $s \cup \{\varphi\} \in S$.

(C3) ($\bigwedge$ rule) If $\bigwedge\Phi \in s$, then for all $\varphi \in \Phi$, $s \cup \{\varphi\} \in S$.

(C4) ($\forall$ rule) If $\forall v\varphi(v) \in s$ and $c \in C$, then $s \cup \{\varphi(c)\} \in S$.

(C5) ($\bigvee$ rule) If $\bigvee\Phi \in s$, then for some $\varphi \in \Phi$, $s \cup \{\varphi\} \in S$.

(C6) ($\exists$ rule) If $\exists v\varphi(v) \in s$, then for some $c \in C$, $s \cup \{\varphi(c)\} \in S$.

(C7) (Equality rules)

(a) If $c,d \in C$ and $c = d \in s$, then $s \cup \{d = c\} \in S$.

(b) If $t$ is a basic term of $M$ w.r.t. $C$, $c \in C$ and $c = t$, $\varphi(t) \in s$, then $s \cup \{\varphi(c)\} \in S$.

(c) If $t$ is a basic term of $M$ w.r.t. $C$, then for some $c \in C$, $s \cup \{c = t\} \in S$.

Clearly the set of all countable sets $s$ of sentences of $M_{\omega_1\omega}$ such that only finitely many $c \in C$ occur in $s$ and $s$ has a model is an ordinary consistency property. Similarly if $\kappa$ is regular it will be clear that the set of all sets $s$ of sentences of $M_{\kappa\omega}$ such that $|s| < \kappa$ and $s$ has a model is a $\kappa$ consistency property. If $\kappa$ is singular with cofinality $\omega$ the corresponding $\kappa$ consistency property is the set of all sets $s$ of sentences of $M_{\kappa\omega}$ such that $|s| < \kappa$, all the constants from $C$ appearing in $s$ come from some $C_n$ and $s$ has a model.

(*ii*) Suppose $\kappa$ is regular and $S$ is a set of sets of sentences of $M_{\kappa\omega}$. $S$ is a $\kappa$ *consistency property* iff for each $s \in S$, $|s| < \kappa$, and all of the following hold:

(C1) If $\varphi$ is a sentence of $M_{\kappa\omega}$, either $\varphi \notin s$ or $\neg\varphi \notin s$.

(C2) If $\{\neg\varphi \mid \varphi \in \Phi\} \subseteq s$, then $s \cup \{\varphi\neg \mid \varphi \in \Phi\} \in S$.

(C3) If $\{\wedge\Phi_i \mid i \in I\} \subseteq s$, then $s \cup \bigcup\{\Phi_i \mid i \in I\} \in S$.

(C4) If $\{\forall v_i \varphi_i(v_i) \mid i \in I\} \subseteq s$ and $C' \subseteq C$, $|C'| < \kappa$, then $s \cup \{\varphi_i(c) \mid c \in C', i \in I\} \in S$.

(C5) If $\{\vee\Phi_i \mid i \in I\} \subseteq s$, then for some $f \in \times\{\Phi_i \mid i \in I\}$, $s \cup \{f(i) \mid i \in I\} \in S$.

(C6) If $\{\exists v_i \varphi_i(v_i) \mid i \in I\} \subseteq s$, then for some $\{c_i \mid i \in I\} \subseteq C$, $s \cup \{\varphi_i(c_i) \mid i \in I\} \in S$.

(C7a) If $\{c_i, d_i \mid i \in I\} \subseteq C$ and $\{c_i = d_i \mid i \in I\} \subseteq s$, then $s \cup \{d_i = c_i \mid i \in I\} \in S$.

(C7b) If $\{t_i \mid i \in I\}$ is a set of basic terms of $M$ w.r.t. $C$, $\{c_i \mid i \in I\} \subseteq C$ and $\{\varphi_i(t_i) \mid i \in I\} \cup \{c_i = t_i \mid i \in I\} \subseteq s$, then $s \cup \{\varphi_i(c_i) \mid i \in I\} \in S$.

(C7c) If $|I| < \kappa$ and $\{t_i \mid i \in I\}$ is a set of basic terms of $M$ w.r.t. $C$, then for some $\{c_i \mid i \in I\} \subseteq C$, $s \cup \{c_i = t_i \mid i \in I\} \in S$.

(*iii*) Suppose $\kappa$ is singular of cofinality $\omega$ and $S$ is a set of sets of sentences of $M_{\kappa\omega}$. $S$ is a $\kappa$ *consistency property* iff for each $s \in S$, $|s| < \kappa$, all the constants from $C$ appearing in $s$ come from some $C_n$, and all of the following hold:

(C1) If $\varphi$ is a sentence of $M_{\kappa\omega}$ then either $\varphi \notin s$ or $\neg\varphi \notin s$.

(C2) If $\{\neg\varphi \mid \varphi \in \Phi\} \subseteq s$, then $s \cup \{\varphi\neg \mid \varphi \in \Phi\} \in S$.

(C3) If $\{\wedge\Phi_i \mid i \in I\} \subseteq s$ and there is a $\lambda < \kappa$ such that for all $i \in I$, $|\Phi_i| < \lambda$, then $s \cup \bigcup\{\Phi_i \mid i \in I\} \in S$.

(C4) If $\{\forall v_i \varphi_i(v_i) \mid i \in I\} \subseteq s$ and $n \in \omega$, then $s \cup \{\varphi_i(c) \mid i \in I, c \in C_n\} \in S$.

(C5) If $\{\vee\Phi_i \mid i \in I\} \subseteq s$ and there is a $\lambda < \kappa$ such that for all $i \in I$, $|\Phi_i| < \lambda$, then there is an $f \in \times\{\Phi_i \mid i \in I\}$ such that $s \cup \{f(i) \mid i \in I\} \in S$.

(C6) If $\{\exists v_i \varphi_i(v_i) \mid i \in I\} \subseteq s$, then there is an $n \in \omega$ and a set $\{c_i \mid i \in I\} \subseteq C_n$ such that $s \cup \{\varphi_i(c_i) \mid i \in I\} \in S$.

(C7) Suppose $n \in \omega$.

(a) If $\{c_i, d_i \mid i \in I\} \subseteq C_n$ and $\{c_i = d_i \mid i \in I\} \subseteq s$, then $s \cup \{d_i = c_i \mid i \in I\} \in S$.

(b) If $\{t_i \mid i \in I\}$ is a set of basic terms of $M$ w.r.t. $C_n$, $\{c_i \mid i \in I\} \subseteq C_n$ and $\{\varphi_i(t_i) \mid i \in I\} \cup \{c_i = t_i \mid i \in I\} \subseteq s$, then $s \cup \{\varphi_i(c_i) \mid i \in I\} \in S$.

(c) If $|I| < \kappa$ and $\{t_i \mid i \in I\}$ is a set of basic terms of $M$ w.r.t. $C_n$, then for some $m \in \omega$ and some $\{c_i \mid i \in I\} \subseteq C_m$, $s \cup \{c_i = t_i \mid i \in I\} \in S$.

1.2 THEOREM. *Every element of an ordinary or a* $\kappa$ *consistency property has a model.*

PROOF: In each case we construct, from a given element $s$ of a consistency property $S$, a countable sequence of elements of $S$ which is in some sense "complete." The constructions for both types of $\kappa$ consistency properties are essentially the same but differ from that for ordinary consistency properties since in that case we can countably list all formulas relevant to the construction. However once we have constructed these sequences a single definition of the model will suffice.

(*i*) Suppose $s_0$ is an element of an ordinary consistency property $S$. We assume, without loss of generality, that every subset of an element of $S$ belongs to $S$. Let $X$ be the set of all sentences in the least set of formulas which contains $s_0$ and all sentences $c = d$ for $c, d \in C$, is closed under subformulas and substitution by constants of $C$ for arbitrary terms, and has as elements all sentences $\varphi\neg$ for which $\neg\varphi$ is an element. $X$ is clearly countable, as is the set $T$ of basic terms of $M$ w.r.t. $C$, so that we can write these sets as $X = \{\varphi_0, \varphi_1, \ldots\}$ and $T = \{t_0, t_1, \ldots\}$. We construct our countable sequence $s_0 \subseteq s_1 \subseteq \ldots$ of elements of $S$ as follows.

Assume $s_n \in S$ and choose $s_{n+1} \in S$ using Definition 1.1(*i*) so that

(a) $s_n \subseteq s_{n+1}$.

(b) If $s_n \cup \{\varphi_n\} \in S$, then $\varphi_n \in s_{n+1}$.

(c) If $s_n \cup \{\varphi_n\} \in S$ and $\varphi_n = \bigvee\Psi$, then for some $\psi \in \Psi$, $\psi \in s_{n+1}$.

(d) If $s_n \cup \{\varphi_n\} \in S$ and $\varphi_n = \exists v\psi(v)$, then for some $c \in C$, $\psi(c) \in s_{n+1}$.

(e) For some $c \in C$, $c = t_n \in s_{n+1}$.

Let $s_\omega = \bigcup\{s_n \mid n \in \omega\}$ and define a relation $\sim$ on elements of $C$ so that $c \sim d$ iff $c = d \in s_\omega$. The equality rules (C7) assure us that $\sim$ is an equivalence relation. We let $\mathfrak{A}$ have universe $A = \{c/\sim \mid c \in C\}$ and interpret the symbols of $M$ as follows.

If $d$ is a constant symbol, $d$ is interpreted by $c/\sim$ iff $c = d \in s_\omega$.

If $\underline{f}$ is a function symbol, $\underline{f}(c_1/\sim,\ldots,c_n/\sim) = c/\sim$ iff $\underline{f}(c_1,\ldots,c_n) = c \in s_\omega$.

If $\underline{R}$ is a relation symbol, $\mathfrak{A} \models \underline{R}(x_1,\ldots,x_n)[c_1/\sim,\ldots,c_n/\sim]$ iff $\underline{R}(c_1,\ldots,c_n) \in s_\omega$.

The equality rules assure us that $\mathfrak{A}$ is well-defined and that if $\varphi(t) \in s_\omega$ and $t$ is interpreted in $\mathfrak{A}$ by $c/\sim$, then $\varphi(c) \in s_\omega$. This together with the consistency rule, (C1), assures us that any atomic or negated atomic sentence of $s_\omega$ holds in $\mathfrak{A}$. An induction argument on the complexity of sentences then shows that every sentence of $s_\omega$ holds in $\mathfrak{A}$, so in particular $\mathfrak{A}$ is a model of $s_0$. Note that $\mathfrak{A}$ is at most countable.

*(ii)* Suppose $\kappa$ is regular and $s_0$ is an element of a $\kappa$ consistency property $S$. Assume $s_n \in S$, $C_n$ is the set of

constants of $C$ which appear in $s_n$ and $T_n$ is the set of basic terms of $M$ w.r.t. $C$ which appear in $s_n$. Since $\kappa$ is regular we know $T_n$ and $C_n$ have cardinality less than $\kappa$ for each $n \in \omega$. We now let $s_{n+1} = s_n^8$ where $s_n^1 - s_n^8$ are defined to be elements of $S$ as follows:

$s_n^1 = s_n \cup \{\varphi\neg \mid \neg\varphi \in s_n\}$

$s_n^2 = s_n^1 \cup \bigcup\{\Phi \mid \wedge\Phi \in s_n\}$

$s_n^3 = s_n^2 \cup \{\varphi(c) \mid \forall x_\varphi \varphi(x_\varphi) \in s_n,\ c \in C_n\}$

$s_n^4 = s_n^3 \cup \{f(\Phi) \mid \vee\Phi \in s_n\}$ for an appropriate $f \in X\{\Phi \mid \vee\Phi \in s_n\}$

$s_n^5 = s_n^4 \cup \{\varphi(c_\varphi) \mid \exists x_\varphi \varphi(x_\varphi) \in s_n\}$ for an appropriate

$\{c_\varphi \mid \exists x_\varphi \varphi(x_\varphi) \in s_n\} \subseteq C$

$s_n^6 = s_n^5 \cup \{d = c \mid c = d \in s_n;\ c, d \in C_n\}$

$s_n^7 = s_n^6 \cup \{\varphi(c) \mid \{\varphi(t), c = t\} \subseteq s_n;\ c \in C_n;\ t \in T_n\}$

$s_n^8 = s_n^7 \cup \{c_t = t \mid t \in T_n\}$ for an appropriate $\{c_t \mid t \in T_n\} \subseteq C$.

In this case the model $\mathfrak{A}$, defined above, has cardinality strictly less than $\kappa$ if $\kappa > \omega$ and is at most countable if $\kappa = \omega$.

*(iii)* Suppose $\kappa > \omega$, $cf(\kappa) = \omega$ and $s_0$ is an element of a $\kappa$ consistency property. Let $F_k$ be the set of $k$-place function symbols of $L$ and $B$ the set of constants of $L$ which appear in $s_0$. Since $B$ and $F_k$ may have cardinality $\kappa$, we write them as countable unions $\bigcup\{F_{k,n} \mid n \in \omega\}$ and $\bigcup\{B_n \mid n \in \omega\}$, with $|F_{k,n}| \leq \kappa_n$ and $|B_n| \leq \kappa_n$ where $\kappa = \bigcup\{\kappa_n \mid n \in \omega\}$. For each $n \in \omega$, let $T_n = C_n \cup B_n \cup \{\underline{f}(c_1, \ldots, c_n) \mid \underline{f} \in F_{k,n};\ c_1, \ldots, c_n \in C_n;\ k \in \omega\}$. Each $T_n$ then has cardinality less than $\kappa$. We now let $s_{n+1} = s_n^8$ where we define $s_n^1 - s_n^8$ as we did in *(ii)* except that we consider only conjunctions, $\wedge\Phi$, and disjunctions, $\vee\Phi$, from $s_n$ for which $|\Phi| \leq \kappa_n$.

In this case the model $\mathfrak{A}$ has cardinality at most $\kappa$. ⊣

COROLLARY (Downward Löwenheim-Skolem Theorem). *If* $\varphi$ *is a sentence of* $L_{\kappa\omega}$ *and* $\varphi$ *has a model then* $\varphi$ *has a model of cardinality*

(a) *strictly less than* $\kappa$ *if* $\kappa$ *is regular and* $\kappa > \omega$,

(b) *at most* $\kappa$ *if* $cf(\kappa) = \omega$,

(c) *at most* $\lambda$ *if* $\kappa = \lambda^{+}$ *and* $cf(\lambda) = \omega$.

PROOF: If $\varphi$ has a model then $\{\varphi\}$ is an element of a consistency property. Except for the case $\kappa = \lambda^{+}$, $\lambda$ a singular cardinal of cofinality $\omega$, the models constructed above have the desired cardinalities. In this last case, there is a sentence $\varphi'$ of $L_{\lambda\omega}$ which is equivalent to $\varphi$ and has a model of cardinality at most $\lambda$. ⊣

As in the above corollary we will often get better results for $L_{\lambda^{+}\omega}$, $cf(\lambda) = \omega$, if we use the fact that the expressive power of $L_{\lambda^{+}\omega}$ and $L_{\lambda\omega}$ is the same for singular $\lambda$. To this end we introduce the following notation -- if $\lambda$ is a singular cardinal of cofinality $\omega$ and $\varphi$ is a formula of $L_{\lambda^{+}\omega}$ we let $\bar{\varphi}$ be any formula of $L_{\lambda\omega}$ which is like $\varphi$ except that each occurence of a subformula $\bigwedge\Psi$ ($\bigvee\Psi$), for $|\Psi| = \lambda$, is replaced by $\bigwedge\{\bigwedge\Psi_n \mid n\in\omega\}$ ($\bigvee\{\bigvee\Psi_n \mid n\in\omega\}$) for some $\{\Psi_n \mid n\in\omega\}$ such that $|\Psi_n| < \lambda$ and $\Psi = \bigcup\{\Psi_n \mid n\in\omega\}$.

2. <u>Completeness Theorems</u>. We now introduce the proof systems for which consistency is equivalent to being an element of a consistency property.

<u>Axioms for $\overset{\circ}{L}_{\kappa\omega}$</u>: All formulas of $L_{\kappa\omega}$ which are

1. substitutions of the tautologies in $\neg$ and $\wedge$ used to

axiomatize finitary propositional logic;

2. of the form $(\neg\varphi \longleftrightarrow \varphi\neg)$;

3. of the form $\wedge\Phi \dashrightarrow \varphi$, for any $\varphi \in \Phi$;

4. of the form $\forall x\varphi(x) \longrightarrow \varphi(t)$, where $t$ is a term of $L$ which is free for $x$ in $\varphi(x)$;

5. equality axioms.

Rules of Inference for $L_{\omega_1\omega}$:

1. From $\varphi$ and $\varphi \longrightarrow \psi$ infer $\psi$. (Modus Ponens)

2. From $\psi \longrightarrow \varphi(x)$ infer $\psi \longrightarrow \forall x\varphi(x)$, where $x$ does not occur free in $\psi$. (Generalization)

3. From $\psi \longrightarrow \varphi$, for all $\varphi \in \Phi$, infer $\psi \longrightarrow \wedge\Phi$. (Conjunction)

Rules of Inference for $L_{\kappa\omega}$, $\kappa$ regular:

1. Modus Ponens.

2. From $\psi \longrightarrow \vee\{\varphi_i(x_i) \mid i\in I\}$ infer $\psi \longrightarrow \vee\{\forall x_i\varphi_i(x_i) \mid i\in I\}$, where, for each $i \in I$, $x_i$ does not occur free in $\psi$ or $\varphi_j(x_j)$ for $j \neq i$. (Disjunctive Generalization)

3. From $\psi \longrightarrow \neg\wedge\{f(i) \mid i\in I\}$, for all $f \in \times\{\Phi_i \mid i\in I\}$, infer $\psi \longrightarrow \neg\wedge\{\vee\Phi_i \mid i\in I\}$. (Distributive Law)

Note that the Conjunction Rule is a special case of the Distributive Law where $I = \{i\}$ and $\Phi_i = \{\neg\varphi \mid \varphi\in\Phi\}$.

Rules of Inference for $L_{\kappa\omega}$, $\kappa > \omega$ and $cf(\kappa) = \omega$.

1. Modus Ponens.

2. Disjunctive Generalization.

3. If there is a $\lambda < \kappa$ such that for all $i \in I$, $|\Phi_i| < \lambda$, then

from $\psi \to \neg\bigwedge\{f(i) \mid i \in I\}$, for all $f \in X\{\Phi_i \mid i \in I\}$, infer $\psi \to \neg\bigwedge\{\bigvee\Phi_i \mid i \in I\}$. ($\kappa_n$ Distributive Laws)

We now let $\vdash_{L_{\kappa\omega}} \varphi$ mean that $\varphi$ is a theorem of $L_{\kappa\omega}$, where the class of theorems is the least class of formulas of $L_{\kappa\omega}$ containing all the axioms and closed under the rules of inference. Clearly we have $\vdash_{L_{\kappa\omega}} \varphi$ iff there is a sequence of formulas of $L_{\kappa\omega}$, $\langle\varphi_\zeta \mid \zeta \leq \lambda\rangle$, such that $\varphi_\lambda = \varphi$ and for all $\zeta \leq \lambda$, $\varphi_\zeta$ is an axiom or is inferred from earlier formulas by a rule of inference. If $\kappa = \omega_1$ then the sequence is at most countable, otherwise it has length less than $2^{*\kappa}$. We will call such a sequence a proof of $\varphi$. We can similarly define a proof system for $M_{\kappa\omega}$ and let $\vdash_{M_{\kappa\omega}} \varphi$ mean that $\varphi$ is a theorem of $M_{\kappa\omega}$. If $\varphi$ is a formula of $L_{\kappa\omega}$ and is a theorem of $M_{\kappa\omega}$, we can replace each occurrence of a constant $c \in C$ by a variable which doesn't appear in the proof in $M_{\kappa\omega}$ and thereby construct a proof in $L_{\kappa\omega}$. Thus for $\varphi \in L_{\kappa\omega}$ we have $\varphi$ is consistent w.r.t. $\vdash_{L_{\kappa\omega}}$ iff $\varphi$ is consistent w.r.t. $\vdash_{M_{\kappa\omega}}$.

2.1 THEOREM. *Suppose* $\Phi$ *is a set of sentences of* $M_{\kappa\omega}$ *and*

(*i*) $\kappa = \omega_1$, $\Phi$ *is finite, and only finitely many* $c \in C$ *occur in* $\Phi$,

(*ii*) $\kappa$ *is regular and* $|\Phi| < \kappa$, *or*

(*iii*) $\kappa$ *is singular of cofinality* $\omega$, $|\Phi| < \kappa$ *and all the constants from* $C$ *appearing in* $\Phi$ *come from some* $C_n$,

*then* $\bigwedge\Phi$ *is consistent w.r.t.* $\vdash_{M_{\kappa\omega}}$ *iff there is a consistency property* $S$ *such that* $\Phi \in S$.

PROOF: Clearly our proof system is correct. Also if $S$ is a consistency property and $\Phi \in S$ then $\bigwedge\Phi$ has a model. Hence we have $\neg\bigwedge\Phi$ is not valid and therefore $\neg\bigwedge\Phi$ is not a theorem of

$M_{\kappa\omega}$ i.e. $\wedge\Phi$ is consistent w.r.t. $\vdash_{M_{\kappa\omega}}$.

We claim that S is a consistency property if S is the set of all sets s of sentences of $M_{\kappa\omega}$ such that $\wedge s$ is consistent w.r.t. $\vdash_{M_{\kappa\omega}}$ and

(*i*) $\kappa = \omega_1$, s is finite and only finitely many $c \in C$ occur in s,

(*ii*) $\kappa$ is regular and $|s| < \kappa$, or

(*iii*) $\kappa$ is singular of cofinality $\omega$, $|s| < \kappa$ and all constants from C appearing in s come from some $C_n$.

Each of the rules C1 - C7 follows directly from the axioms, simple applications of the rules of inference and the fact that if $s' \subseteq s$ and $\vdash_{M_{\kappa\omega}} \wedge s \rightarrow \neg\wedge s'$, then $s \notin S$. To illustrate this we check (C7c) for $\kappa$ regular.

Suppose $\{t_i \mid i \in I\}$ is a set of less than $\kappa$ basic terms of M w.r.t. C. Choose a set of I distinct constants, $\{c_i \mid i \in I\} \subseteq C$, which do not appear in s and assume $s \cup \{c_i = t_i \mid i \in I\} \notin S$. Then $\vdash_{M_{\kappa\omega}} \wedge s \rightarrow \neg\wedge\{c_i = t_i \mid i \in I\}$, i.e. $\vdash_{M_{\kappa\omega}} \wedge s \rightarrow \vee\{c_i \neq t_i \mid i \in I\}$. We can clearly find a set of $|I|$ distinct variables $\{w_i \mid i \in I\}$ and a proof of the last sentence in which no $w_i$ appears. If in this proof we replace each $c_i$ by $w_i$ we then have $\vdash_{M_{\kappa\omega}} \wedge s \rightarrow \vee\{w_i \neq t_i \mid i \in I\}$ to which we apply rule 2 to get $\vdash_{M_{\kappa\omega}} \wedge s \rightarrow \vee\{\forall w_i (w_i \neq t_i) \mid i \in I\}$. Using axioms three and four we get $\vdash_{M_{\kappa\omega}} \wedge s \rightarrow \vee\{t_i \neq t_i \mid i \in I\}$ so that

$$\vdash_{M_{\kappa\omega}} \wedge s \rightarrow (\wedge\{t_i = t_i \mid i \in I\} \wedge \neg\wedge\{t_i = t_i \mid i \in I\})$$

and hence $s \notin S$.

Hence if $\wedge\Phi$ is consistent w.r.t. $\vdash_{M_{\kappa\omega}}$ we have $\Phi$ an element of the consistency property S. ⊣

2.2 COROLLARY (Completeness Theorem, Karp [14]). (a) *If* $\varphi$ *is a sentence of* $L_{\kappa\omega}$ *then* $\vdash_{M_{\kappa\omega}} \varphi$ *iff* $\models \varphi$.

(b) *If* $\kappa$ *is singular of cofinality* $\omega$ *and* $\varphi$ *is a sentence of* $L_{\kappa^+\omega}$ *then* $\vdash_{L_{\kappa\omega}} \overline{\varphi}$ *iff* $\models \varphi$.

PROOF: (a) $\varphi$ is not a theorem of $L_{\kappa\omega}$ iff $\neg\varphi$ is consistent w.r.t. $\vdash_{L_{\kappa\omega}}$ iff $\neg\varphi$ is consistent w.r.t. $\vdash_{M_{\kappa\omega}}$ iff $\{\neg\varphi\}$ is an element of a consistency property iff $\{\neg\varphi\}$ has a model iff $\varphi$ is not valid.

(b) Part (a) shows $\vdash_{L_{\kappa\omega}} \overline{\varphi}$ iff $\models\overline{\varphi}$ and clearly $\models\overline{\varphi}$ iff $\models\varphi$. ⊣

A proof similar to that of Theorem 2.1 shows that for $\kappa$ regular, the validity predicate for $L_{\kappa\omega}$ is $\Sigma_1$ in the power set operation on $H_{2^{*}\kappa}$ if the symbols of $L$ are in $H_{2^{*}\kappa}$. If $\kappa$ is a strong limit cardinal of cofinality $\omega$, validity is $\Sigma_1$ rather than $\Sigma_1$ in the power set, i.e. if the symbols of $L$ come from $H_{\kappa^+}$ then validity is $\Sigma_1$ on $H_{\kappa^+}$. It is known that assuming the Generalized Continuum Hypothesis, validity is $\Sigma_1$ on $H_{\kappa^+}$ iff $cf(\kappa) = \omega$, [15].

3. <u>Interpolation Theorems</u>. For all the languages we are considering there are interpolation theorems which generalize theorems of $L_{\omega\omega}$ to give infinitary versions of preservation and definability theorems. If however $\kappa > \omega_1$, these results involve languages with infinite quantifiers, i.e. the interpolating sentences will be in $L_{2^{*}\kappa\kappa}$. As usual we consider sentences of $L_{\kappa^+\omega}$, $\kappa$ singular of cofinality $\omega$, as sentences of $L_{\kappa\omega}$ to find interpolating sentences in $L_{2^{*}\kappa\kappa}$ instead of $L_{(2^{\kappa})^{+}\kappa^{+}}$.

The Craig interpolation theorem was generalized to $L_{\omega_1\omega}$ by Lopez-Escobar [20] and then to $L_{\kappa\omega}$, $\kappa$ regular, by Malitz [23].

Chang [4] showed that, for $\kappa$ an uncountable strong limit cardinal of cofinality $\omega$, the interpolation sentence for $L_{\kappa^+\omega}$ could be found in $L_{\kappa^+\kappa}$ instead of $L_{(2^\kappa)^+\kappa^+}$. Lopez-Escobar also generalized the stronger Lyndon interpolation theorem to $L_{\omega_1\omega}$ and Malitz extended the Łoś - Tarski results on preservation of submodels to $L_{\omega_1\omega}$ using an interpolation theorem. These theorems were then extended by Chang to the case $L_{\kappa^+\omega}$, $\kappa$ an uncountable strong limit cardinal of cofinality $\omega$.

Although we are considering infinitary formulas it is clear that in any formula every occurrence of a symbol is within the scope of only a finite number of occurrences of $\neg$. An occurrence of a symbol is called *positive* if it lies within the scope of an even number of negation signs. Otherwise it is called a *negative* occurrence. A sentence $\varphi$ is called *positive* iff every occurrence of a relation, function, or constant symbol or of the equality symbol in $\varphi$ is positive. A formula $\varphi$ is called *universal* if every occurrence of the quantifier $\forall$ in $\varphi$ is positive and every occurrence of the quantifier $\exists$ is negative. A formula $\varphi$ is called *existential* iff $\neg\varphi$ is universal.

3.1 CRAIG - LYNDON INTERPOLATION THEOREM. *Suppose* $\varphi$ *and* $\psi$ *are sentences of* $L_{\kappa\omega}$ *such that* $\models\varphi\rightarrow\psi$. *There is a sentence* $\theta$ *of* $L_{\omega_1\omega}$, *if* $\kappa=\omega_1$, *of* $L_{2^{*\kappa}\kappa}$, *if* $\kappa$ *is regular or of cofinality* $\omega$, *such that* $\models\varphi\rightarrow\theta$ *and* $\models\theta\rightarrow\varphi$, *every function and constant symbol which occurs in* $\theta$ *occurs in both* $\varphi$ *and* $\psi$, *and every relation symbol which occurs positively (negatively) in* $\theta$ *occurs positively (negatively) in both* $\varphi$ *and* $\psi$.

PROOF: Assume this theorem is proved for all languages $L'_{\kappa\omega}$ and all sentences $\varphi$ and $\psi$ of $L'_{\kappa\omega}$ such that every constant and function symbol of $L'$ occurs in both $\varphi$ and $\psi$ and every relation symbol of $L'$ occurs in one of them. Suppose now we have an arbitrary language $L_{\kappa\omega}$ and arbitrary sentences $\varphi$ and $\psi$ of $L_{\kappa\omega}$. Let $L'$ be the language which includes as symbols the constant and function symbols of $L$ which appear in both $\varphi$ and $\psi$, the relation symbols of $L$ which appear in either $\varphi$ or $\psi$, and new relation symbols $\underline{R}_d$ and $\underline{R}_{\underline{f}}$ for each constant and function symbol of $L$ which occurs in exactly one of $\varphi$ or $\psi$. Let

$$D_{\varphi-\psi} = \{\text{constants of } L \text{ appearing in } \varphi \text{ and not in } \psi\},$$

and

$$F^n_{\varphi-\psi} = \{n \text{ place function symbols of } L \text{ appearing in } \varphi \text{ and not in } \psi\},$$

and define the sentence $\varphi'$ of $L'_{\kappa\omega}$ as

$$\varphi'' \wedge \bigwedge_{d\in D_{\varphi-\psi}} (\exists! x \underline{R}_d x) \wedge \bigwedge_{n\in\omega} \Bigg( \bigwedge_{\underline{f}\in F^n_{\varphi-\psi}} (\forall x_1 \ldots x_n \exists! x_{n+1} \underline{R}_{\underline{f}} x_1 \ldots x_n x_{n+1}) \Bigg),$$

where $\varphi''$ is defined by replacing, for each $d \in D_{\varphi-\psi}$, each occurrence of an atomic formula $\underline{S}(..d..)$ by $\exists x(\underline{R}_d x \wedge \underline{S}(..x..))$ and for $t = \underline{f}(t_1,\ldots,t_n)$, where $\underline{f} \in F^n_{\varphi-\psi}$ and $t_1 \ldots t_n$ have no occurrences of function symbols or constants from any $F^m_{\varphi-\psi}$ or $D_{\varphi-\psi}$, replacing each occurrence of an atomic formula $\underline{S}(..t..)$ by $\exists x(\underline{R}_{\underline{f}}(t_1,\ldots,t_n,x) \wedge \underline{S}(..x..))$. Define $\psi'$ similarly.

Since $\models \varphi \rightarrow \psi$ we have $\models \varphi' \rightarrow \psi'$ and so under our present assumptions there is a sentence $\theta$ of $L'_{\omega_1\omega}$ or $L'_{2^{*\kappa}\kappa}$ such that $\models \varphi' \rightarrow \theta$ and $\models \theta \rightarrow \psi'$ and every function and constant symbol which occurs in $\theta$ occurs in both $\varphi'$ and $\psi'$ and every relation symbol which occurs positively (negatively) in $\theta$ occurs positively (negatively) in both $\varphi'$ and $\psi'$. No new relation symbols can occur in $\theta$ since no new relation symbols occur in both $\varphi'$ and

$\psi'$. Thus we have $\theta$ is a sentence of $L$ such that $\models \varphi \rightarrow \theta$ and $\models \theta \rightarrow \psi$ and every function and constant symbol which occurs in $\theta$ occurs in both $\varphi$ and $\psi$ and every relation symbol which occurs positively (negatively) in $\theta$ occurs similarly in both $\varphi$ and $\psi$.

We will therefore assume for the remainder of the proof that every function and constant symbol of $L$ appears in both $\varphi$ and $\psi$ and every relation symbol of $L$ appears in one of them.

Suppose $\kappa$ is regular, $\kappa \neq \omega_1$. For any sentence $\sigma$ of $L_{\kappa\omega}$ let $X_\sigma$ be the set of sentences $\sigma'$ of $M_{2^{*\kappa}\kappa}$ such that less than $\kappa$ $c \in C$ occur in $\sigma'$ and every relation symbol which occurs positively (negatively) in $\sigma'$ occurs positively (negatively) in $\sigma$. For other $\kappa$ we need the following modifications in the definition of $X_\sigma$. If $\kappa = \omega_1$, $X_\sigma$ is a set of sentences of $M_{\omega_1\omega}$ and only finitely many $c \in C$ occur in each sentence of $X_\sigma$. If $\kappa > \omega$ and $cf(\kappa) = \omega$, the $c \in C$ which occur in a sentence of $X_\sigma$ must all come from the same $C_n$.

Again suppose $\kappa$ is regular and $\kappa \neq \omega_1$. Let $S$ be the set of all $s \subseteq M_{\kappa\omega}$ such that $|s| < \kappa$ and $s$ can be written as $s = s_1 \cup s_2$ with

(1) $s_1 \subseteq X_\varphi$, $s_2 \subseteq X_{\neg\psi}$ and

(2) there is no $\theta \in X_\varphi \cap X_\psi$ such that $\models \wedge s_1 \rightarrow \theta$ and $\models \wedge s_2 \rightarrow \neg\theta$.

We claim $S$ is a $\kappa$ consistency property. We also claim that the obvious modifications in the definition of $S$ produce an ordinary consistency property if $\kappa = \omega_1$ and a $\kappa$ consistency property if $\kappa > \omega$ and $cf(\kappa) = \omega$. We present only the proof for $\kappa$ regular, $\kappa \neq \omega_1$. Assume $s \in S$.

(C1) Suppose $\{\sigma, \neg\sigma\} \subseteq s$. If $\sigma \in s_1$ and $\neg\sigma \in s_2$ then $\sigma \in X_\varphi \cap X_\psi$ and $\theta = \sigma$ contradicts (2). Similarly if $\neg\sigma \in s_1$ and $\sigma \in s_2$. If the contradictory set is a subset of $s_1$ then

$\theta = \forall x(x \neq x)$ contradicts (2) and if it is a subset of $s_2$ then $\theta = \forall x(x = x)$ contradicts (2).

(C2) Suppose $\{\neg\sigma \mid \sigma\in\Sigma\} \subseteq s$ and let $s_i' = s_i \cup \{\sigma\neg \mid \sigma\in\Sigma,\ \neg\sigma\in s_i\}$. Clearly $s_1'$ and $s_2'$ satisfy (1) and if there is a $\theta \in X_\varphi \cap X_\psi$ such that $\models \wedge s_1' \to \theta$ and $\models \wedge s_2' \to \neg\theta$ then that $\theta$ will contadict (2) for $s$.

(C3), (C4) and (C7a) are as trivial to check as (C2) is.

(C5) Suppose $\{\vee\Sigma_i \mid i\in I\} \subseteq s$. Let $I_k = \{i\in I \mid \vee\Sigma_i \in s_k\}$. Suppose for all $f \in X\{\Sigma_i \mid i\in I_1\}$, $s_f = s \cup \{f(i) \mid i\in I_1\}$ and $s_f \notin S$. Let $s_{f,1} = s_1 \cup \{f(i) \mid i\in I_1\}$ and $s_{f,2} = s_2$. Then for all $f \in X\{\Sigma_i \mid i\in I_1\}$, there are $\theta_f \in X_\varphi \cap X_\psi$ such that $\models (\wedge s_1 \wedge \wedge\{f(i) \mid i\in I_1\}) \to \theta_f$ and $\models \wedge s_2 \to \neg\theta_f$. Let $\theta = \vee\{\theta_f \mid f\in X\{\Sigma_i \mid i\in I_1\}\}$. Clearly $\theta \in X_\varphi \cap X_\psi$ and $\models (\wedge s_1 \wedge \vee\{\wedge\{f(i) \mid i\in I_1\} \mid f\in X\{\Sigma_i \mid i\in I_1\}\}) \to \theta$ and $\models \wedge s_2 \to \neg\theta$. But since the distributive law is valid $\models \wedge s_1 \to \theta$, contradicting $s \in S$. Using the same argument we get there is a $g \in X\{\Sigma_i \mid i\in I_2\}$ such that $s \cup \{f(i) \mid i\in I_1\} \cup \{g(i) \mid i\in I_2\} \in S$. So if $h = f \cup g$, $h \in X\{\Sigma_i \mid i\in I\}$ and $s \cup \{h(i) \mid i\in I\} \in S$. Note that here is where our conjunctions may get large, i.e. $|X\{\Sigma_i \mid i\in I\}| < 2^{*\kappa}$ but not necessarily less than $\kappa$ unless $\kappa$ is strongly inaccessible.

(C6) Suppose $\{\exists v_i\sigma_i(v_i) \mid i\in I\} \subseteq s$ and $\{c_i \mid i\in I\}$ is a set of $|I|$ distinct constants of $C$ none of which appear in $s$. Let $I_k = \{i\in I \mid \exists v_i\sigma_i(v_i) \in s_k\}$ and $s_1' = s_1 \cup \{\sigma_i(c_i) \mid i\in I_i\}$, $s_2' = s_2$. Suppose $\theta \in X_\varphi \cap X_\psi$ such that $\models (\wedge s_1 \wedge \wedge\{\sigma_i(c_i) \mid i\in I_1\}) \to \theta$ and $\models \wedge s_2 \to \neg\theta$. For $i \in I$, $c_i$ may appear in $\theta$ but not in $s$, hence we may replace all occurrences of $c_i$ by variables and quantify to get $\models (\wedge s_1 \wedge \wedge\{\exists v_i\sigma_i(v_i) \mid i\in I_1\}) \to \exists\{v_i \mid i\in I_1\}\theta(\ldots c_i/v_i \ldots)$ and $\models \wedge s_2 \to \forall\{v_i \mid i\in I_1\}\neg\theta(\ldots c_i/v_i \ldots)$. Thus also $\models \wedge s_1 \to \exists\{v_i \mid i\in I_1\}\theta(\ldots c_i/v_i \ldots)$ contradicting (2) for $s$. To add

$\{\sigma_i(c_i) \mid i \in I_2\}$ to $s_2$ we use the same argument to show that if the new set is not in $S$ there is an appropriate $\theta$ such that $\models \wedge s_1 \rightarrow \forall\{v_i \mid i \in I_2\}\theta(\ldots c_i/v_i \ldots)$ and $\models \wedge s_2 \rightarrow \exists\{v_i \mid i \in I_2\}\neg\theta(\ldots c_i/v_i \ldots)$.

(C7b) Suppose $\{\sigma_i(t_i) \mid i \in I\} \cup \{c_i = t_i \mid i \in I\} \subseteq s$ where $\{c_i \mid i \in I\} \subseteq C$ and $\{t_i \mid i \in I\}$ is a set of basic terms of $M$ w.r.t. $C$. Let $I_j = \{i \in I \mid \sigma_i(t_i) \in s_j\}$, $I_{j,k} = \{i \in I_j \mid c_i = t_i \in s_k\}$ and $s_j' = s_j \cup \{\sigma_i(c_i) \mid i \in I_j\}$. Clearly $s_1'$ and $s_2'$ satisfy (1) and $\models \wedge s_j \rightarrow \wedge\{\sigma_i(c_i) \mid i \in I_{j,j}\}$. Suppose $\theta \in X_\varphi \cap X_\psi$ such that $\models \wedge s_1' \rightarrow \theta$ and $\models \wedge s_2' \rightarrow \neg\theta$. Then $\models (\wedge s_1 \wedge \wedge\{\sigma_i(c_i) \mid i \in I_{1,2}\}) \rightarrow \theta$ and since

$$\models \wedge s_1 \rightarrow (\wedge\{c_i = t_i \mid i \in I_{1,2}\} \rightarrow \wedge\{\sigma_i(c_i) \mid i \in I_{1,2}\}),$$

$$\models \wedge s_1 \rightarrow (\wedge\{c_i = t_i \mid i \in I_{1,2}\} \rightarrow \theta).$$

Hence

$$\models \wedge s_1 \rightarrow (\wedge\{c_i = t_i \mid i \in I_{2,1}\} \wedge (\wedge\{c_i = t_i \mid i \in I_{1,2}\} \rightarrow \theta)).$$

Similarly

$$\models \wedge s_2 \rightarrow (\wedge\{c_i = t_i \mid i \in I_{1,2}\} \wedge (\wedge\{c_i = t_i \mid i \in I_{2,1}\} \rightarrow \neg\theta))$$

so that

$$\models \wedge s_2 \rightarrow (\wedge\{c_i = t_i \mid i \in I_{2,1}\} \rightarrow (\wedge\{c_i = t_i \mid i \in I_{1,2}\} \wedge \neg\theta)).$$

Hence if

$$\theta' = \wedge\{c_i = t_i \mid i \in I_{2,1}\} \wedge (\wedge\{c_i = t_i \mid i \in I_{1,2}\} \rightarrow \theta),$$

$\models \wedge s_1 \rightarrow \theta'$ and $\models \wedge s_2 \rightarrow \neg\theta'$. $\theta' \in X_\varphi \cap X_\psi$ since the only symbols of $L$ which occur in $\theta'$ which might not occur in $\theta$ are function or constant symbols.

(C7c) Suppose $\{t_i \mid i \in I\}$ is a set of less than $\kappa$ basic terms of $M$ w.r.t. $C$. Let $\{c_i \mid i \in I\}$ be a set of $|I|$ distinct elements of $C$ which do not appear in $s$ or $\{t_i \mid i \in I\}$. Each $c_i = t_i$ is an element of $X_\varphi$ and $X_{\neg\psi}$ and hence can be added to either $s_1$

or $s_2$. Let $s_1' = s_1, s_2' = s_2 \cup \{c_i = t_i \mid i \in I\}$ and suppose $\theta \in X_\varphi \cap X_\psi$ such that $\models \wedge s_1' \rightarrow \theta$ and $\models \wedge s_2' \rightarrow \neg\theta$, i.e. $\models \wedge s_1 \rightarrow \theta$ and $\models (\wedge s_2 \wedge \wedge\{c_i = t_i \mid i \in I\}) \rightarrow \neg\theta$. Hence $\models \wedge s_1 \rightarrow \forall\{v_i \mid i \in I\}\theta(\ldots c_i/v_i \ldots)$ and $\models \wedge s_2 \rightarrow \exists\{v_i \mid i \in I\}\neg\theta(\ldots c_i/v_i \ldots)$ contradicting $s \in S$.

Since $S$ is a $\kappa$ consistency property every element of $S$ has a model. But $\models \varphi \rightarrow \psi$ so $\{\varphi, \neg\psi\}$ has no model and therefore is not an element of S. Hence $\varphi \in X_\varphi$ and $\neg\psi \in X_{\neg\psi}$ implies there is a $\theta \in X_\varphi \cap X_\psi$ such that $\models \varphi \rightarrow \theta$ and $\models \neg\psi \rightarrow \neg\theta$. By quantifying out the constants of $C$ which appear in $\theta$ we get the desired interpolating sentence of $L_{2^{*\kappa}\kappa}$. $\dashv$

Note that if $L$ has no function or constant symbols and we assume every sentence of $s_1$ is universal we can conclude $\theta$ is universal since the only time we add an existensial quantifier to $\theta$ is for the existential formulas of $s_1$.

If $\kappa$ is singular of cofinality $\omega$ and $\varphi, \psi \in L_{\kappa^+\omega}$ then the interpolating sentence can be found in $L_{2^{*\kappa}\kappa}$. Hence if $\kappa$ is a strong limit cardinal of cofinality $\omega$ the interpolating sentence will be in $L_{\kappa^+\kappa}$.

It is also clear that if $\varphi(x_1,\ldots,x_n)$ and $\psi(x_1,\ldots,x_n)$ are formulas of $L_{\kappa\omega}$ such that $\models \varphi(x_1,\ldots,x_n) \rightarrow \psi(x_1,\ldots,x_n)$ then there is an interpolating formula $\theta(x_1,\ldots,x_n)$ satisfying the conditions of Theorem 3.1.

3.2 BETH'S DEFINABILITY THEOREM. *Let* $\underline{R}$ *and* $\underline{S}$ *be two* $n$ *place relation symbols not in* $L$. *Let* $L'$ *be the language* $L$ *together with* $\underline{R}$. *Suppose* $\varphi$ *is a sentence of* $L'_{\kappa\omega}$ *and* $\varphi'$ *is the sentence formed by replacing each occurrence of* $\underline{R}$ *in* $\varphi$ *by* $\underline{S}$. *If* $\models (\varphi \wedge \varphi') \rightarrow (\underline{R}(x_1,\ldots,x_n) \rightarrow \underline{S}(x_1,\ldots,x_n))$, *i.e.* $\varphi(\underline{R})$ *defines* $\underline{R}$ *implicitly, then there is a formula* $\theta(x_1,\ldots,x_n)$ *of* $L_{2^{*\kappa}\kappa}$ *for*

$\kappa$ *regular or of cofinality* $\omega$ ($L_{\omega_1\omega}$ *for* $\kappa = \omega_1$), *such that* $\models\varphi \rightarrow (\underline{R}(x_1,\ldots,x_n) \longleftrightarrow \theta(x_1,\ldots,x_n))$, *i.e.* $\theta(x_1,\ldots,x_n)$ *defines* $\underline{R}$ *explicitly.*

PROOF: We apply the Craig interpolation theorem to $(\varphi \wedge \underline{R}(x_1,\ldots,x_n)) \rightarrow (\varphi' \rightarrow \underline{S}(x_1,\ldots,x_n))$ to get a formula $\theta(x_1,\ldots,x_n)$ of $L$ such that $\models(\varphi \wedge \underline{R}(x_1,\ldots,x_n)) \rightarrow \theta(x_1,\ldots,x_n)$ and $\models\theta(x_1,\ldots,x_n) \rightarrow (\varphi' \rightarrow \underline{S}(x_1,\ldots,x_n))$. Since neither $\underline{R}$ nor $\underline{S}$ appears in $\theta$ it follows that $\models\theta(x_1,\ldots,x_n) \rightarrow (\varphi \rightarrow \underline{R}(x_1,\ldots,x_n))$. Thus $\models\varphi \rightarrow (\underline{R}(x_1,\ldots,x_n) \longleftrightarrow \theta(x_1,\ldots,x_n))$. $\dashv$

If $L$ and $L'$ are languages and $\underline{R}$ is an $n$ place relation symbol not in $L$ or $L'$, we write Beth($L,L'$) iff there is an implicit definition of $\underline{R}$ in $L(\underline{R})$ implies there is an explicit definition of $\underline{R}$ in $L'$.

The above theorem then becomes Beth($L_{\omega_1\omega}, L_{\omega_1\omega}$) and Beth($L_{\kappa\omega}, L_{2^{*\kappa}\kappa}$) for $\kappa$ regular or of cofinality $\omega$. Hence for any cardinal $\kappa$, Beth($L_{\kappa^+\omega}, L_{(2^\kappa)^+\kappa^+}$). Also if $\kappa$ is singular of cofinality $\omega$ Beth($L_{\kappa^+\omega}, L_{2^{*\kappa}\kappa}$) and therefore if $\kappa$ is a strong limit cardinal of cofinality $\omega$, Beth($L_{\kappa^+\omega}, L_{\kappa^+\kappa}$). There have also been negative results along this line. Malitz [23] showed not-Beth($L_{\omega_1\omega_1}, L_{\infty\infty}$) and hence not-Beth($L_{\infty\infty}, L_{\infty\infty}$). Gregory [11] showed not-Beth($L_{\infty\omega}, L_{\infty\omega}$) and for every cardinal $\lambda$ not-Beth($L_{\infty\omega}, L_{\infty\lambda}$). He has also shown that if $\kappa$ is an uncountable regular cardinal we cannot lower the bound on the quantifier for $L_{\kappa^+\omega}$, i.e. not-Beth($L_{\kappa^+\omega}, L_{\infty\kappa}$). Friedman [8] has shown that if $\kappa$ is a cardinal with $cf(\kappa) > \omega$ then we cannot lower the bound on the size of the conjunctions to $\kappa^+$, i.e. not-Beth($L_{\kappa^+\omega}, L_{\kappa^+\kappa^+}$).

3.3 Preservation Under Homomorphisms. A sentence $\varphi$ is said to be *preserved under homomorphisms relative to* $\sigma$ iff whenever $\mathfrak{A}$ is a model of $\sigma \wedge \varphi$ then every homomorphic image of $\mathfrak{A}$ which is a model of $\sigma$ is a model of $\varphi$. The main result of this section is that a sentence is preserved under homomorphisms (relative to $\sigma$) iff there is a positive sentence, of an appropriate language, to which it is equivalent (in every model of $\sigma$). To show this we need a modified version of the model existence theorem and the Lyndon interpolation theorem.

MODEL EXISTENCE THEOREM WITHOUT EQUALITY. *Suppose* L *has no function or constant symbols and* S *satisfies the definition of a consistency property* (1.1) *except for the equality rules. If the equality symbol does not occur in any element of* S *then every element of* S *has a model.*

PROOF: For $\kappa \neq \omega_1$ we construct the sequence $s_n \in S$ as in Theorem 1.2 except we let $s_{n+1} = s_n^5$. For $\kappa = \omega_1$, we again construct the sequence as in Theorem 1.2 except we list only the sentences in which the equality symbol doesn't appear and we do not include in $s_{n+1}$ any sentence $c = t_n$. The model on constants (rather than on equivalence classes of constants) is clearly a model of s. ⊣

LYNDON INTERPOLATION THEOREM WITHOUT EQUALITY. *Suppose* L *has no function or constant symbols,* $\varphi$ *and* $\psi$ *are sentences of* $L_{\kappa\omega}$ *in which the equality symbol doesn't occur,* $\vDash \varphi \rightarrow \psi$ *and* $\varphi$ *and* $\neg\psi$ *both have models. If* $\kappa$ *is regular or has cofinality* $\omega$, *there is a sentence* $\theta$ *of* $L_{2^{*}\kappa\,\kappa}$ *(*$L_{\omega_1\omega}$ *if* $\kappa = \omega_1$*) in which the equality symbol doesn't occur and such that* $\vDash \varphi \rightarrow \theta$ *and* $\vDash \theta \rightarrow \psi$ *and every relation symbol which occurs positively (negatively) in* $\theta$ *occurs positively (negatively) in both* $\varphi$ *and* $\psi$.

PROOF: This proof is essentially the same as that of Theorem

3.1. We let $X_\sigma$ be as in the proof of that theorem and define $Y_\sigma$ as the set of all $\sigma' \in X_\sigma$ in which the equality symbol doesn't appear. We define S as before except that we replace $X_\sigma$ by $Y_\sigma$ throughout and assume $s_1$ and $s_2$ both have models. We can use the model existence theorem without equality and therefore only check (C1) - (C6) of Definition 1.1. But clearly (C2) - (C6) work exactly as in Theorem 3.1, as does (C1) except when $\sigma, \neg\sigma \in s_j$. But we are assuming that each $s_j$ has a model so this case cannot occur. Thus again we know $\{\varphi, \neg\psi\}$ has no model implies $\{\varphi, \neg\psi\} \notin S$ and since $\varphi$ and $\neg\psi$ have models we get the desired $\theta$. ⊣

CORROLLARY. *Let* $\sigma$ *and* $\varphi$ *be sentences of* $L_{\kappa\omega}$ *such that* $\varphi$ *has a model.* $\varphi$ *is preserved under homomorphisms relative to* $\sigma$ *iff there is a positive sentence* $\theta$ *of* $L_{2^{*\kappa}\kappa}$, *for* $\kappa$ *regular or of cofinality* $\omega$, *of* $L_{\omega_1\omega}$ *for* $\kappa = \omega_1$, *such that* $\models \sigma \to (\varphi \longleftrightarrow \theta)$.

PROOF: If $\varphi$ is preserved under homomorphisms relative to $\sigma$ we eliminate all function and constant symbols of $\sigma$ and $\varphi$ as we did in the proof of Theorem 3.1 and then apply the Lyndon interpolation theorem without equality to

$$(C \wedge \sigma(\underline{E}) \wedge \varphi(\underline{E})) \longrightarrow ((H \wedge C' \wedge \sigma'(\underline{E}')) \longrightarrow \varphi'(\underline{E}'))$$

where $\underline{E}$ is a new binary relation which replaces each occurrence of the equality symbol in $\sigma$ and $\varphi$ to form $\sigma(\underline{E})$ and $\varphi(\underline{E})$; C is the sentence stating $\underline{E}$ is an equivalence and a congruence relation w.r.t. each relation symbol in $\sigma$ and $\varphi$; $\underline{R}'$ is a new relation symbol for each symbol $\underline{R}$ of $\sigma(\underline{E})$ and $\varphi(\underline{E})$, and C', $\sigma'(\underline{E}')$ and $\varphi'(\underline{E}')$ are formed by replacing each $\underline{R}$ by $\underline{R}'$; and H is the conjunction of all sentences $\forall x_1 \ldots x_n (\underline{R}(x_1, \ldots, x_n) \longrightarrow \underline{R}'(x_1, \ldots, x_n))$ for the relation symbols $\underline{R}$ of $\sigma(\underline{E})$ and $\varphi(\underline{E})$. Thus there is a sentence $\theta(\underline{E})$ involving only $\underline{E}$ and the relation symbols of L such that $\models (C \wedge \sigma(\underline{E}) \wedge \varphi(\underline{E})) \longrightarrow \theta(\underline{E})$ and $\models \theta(\underline{E}) \longrightarrow ((H \wedge C' \wedge \sigma'(\underline{E}')) \longrightarrow \varphi'(\underline{E}'))$. Since the relation symbols of $\theta$ appear only positively in $(H \wedge C' \wedge \sigma'(\underline{E}')) \rightarrow \varphi'(\underline{E}')$, $\theta$ is a positive sentence. Replacing the new relation symbols by the old we get $\models \theta(\underline{E}) \longrightarrow ((C \wedge \sigma(\underline{E})) \longrightarrow \varphi(\underline{E}))$ so that $\models (C \wedge \sigma(\underline{E})) \longrightarrow (\varphi(\underline{E}) \longleftrightarrow \theta(\underline{E}))$. We get the desired result by replacing $\underline{E}$ by the equality symbol.

Clearly every positive sentence is preserved under homomorphisms, hence if $\models \sigma \rightarrow (\varphi \longleftrightarrow \theta)$ for a positive sentence $\theta$, $\varphi$ is preserved under homomorphisms relative to $\sigma$. ⊣

Clearly we can state this corollary for $L_{\kappa^+\omega}$, $\kappa$ singular of cofinality $\omega$, and find the positive sentence in $L_{2^{*\kappa}\kappa}$ or in $L_{\kappa^+\kappa}$ if $\kappa$ is also a strong limit cardinal.

3.4 MALITZ INTERPOLATION THEOREM. *Suppose* L *has no function symbols. If* $\varphi$ *and* $\psi$ *are sentences of* $L_{\kappa\omega}$ *such that* $\psi$ *is universal and* $\models \varphi \longrightarrow \psi$ *then there is a universal sentence* $\theta$ *of* $L_{2^{*\kappa}\kappa}$ *if* $\kappa$ *is regular or of cofinality* $\omega$ *(*$L_{\omega_1\omega}$ *if* $\kappa = \omega_1$*), such that* $\models \varphi \longrightarrow \theta$, $\models \theta \longrightarrow \psi$ *and every relation and constant symbol which appears in* $\theta$ *appears in both* $\varphi$ *and* $\psi$.

PROOF: We can assume $\psi$ is quantifier free since the bound variables in $\psi$ can be replaced by new constants to form a sentence $\psi'$ for which $\models \varphi \longrightarrow \psi$ implies $\models \varphi \longrightarrow \psi'$ and $\models \theta \longrightarrow \psi'$ implies $\models \theta \longrightarrow \psi$. We present only the proof for $\kappa$ regular, the other cases working similarly.

Let S be the set of all sets of sentences of $M_{\kappa\omega}$ such that $|s| < \kappa$ and s can be written as $s = s_1 \cup s_2$ where all sentences in $s_2$ are quantifier free and such that there is no universal sentence $\theta$ of $M_{2^{*\kappa}\kappa}$ such that:

(1) $\models \wedge s_1 \rightarrow \theta$, $\models \wedge s_2 \rightarrow \neg\theta$ and

(2) every relation and constant symbol of $M$ which occurs in $\theta$ occurs in both $s_1$ and $s_2$.

We claim $S$ is a $\kappa$ consistency property. Assume $s \in S$ and $s = s_1 \cup s_2$ satisfies the above conditions. We check only those rules whose proofs differ from that in Theorem 3.1.

(C4) Suppose $\{\forall v_i \sigma_i(v_i) \mid i \in I\} \subseteq s$, $C' \subseteq C$, $|C'| < \kappa$. Since all formulas in $s_2$ are quantifier free, $\{\forall v_i \sigma_i(v_i) \mid i \in I\} \subseteq s_1$. Let $s_1' = s_1 \cup \{\sigma_i(c) \mid i \in I, c \in C'\}$ and $s_2' = s_2$. Suppose $s' \notin S$ so that there is a $\theta$ which satisfies (1) and (2) for $s_1'$ and $s_2'$. Every relation and constant symbol of $L$ which occurs in $\theta$ occurs in $s_1$ and $s_2$ and every $c \in C$ which occurs in $\theta$ occurs in $s_2$ but not necessarily in $s_1$. Let $C''$ be the set of $c \in C'$ such that $c$ occurs in $\theta$ but doesn't occur in $s_1$. For each $c \in C''$ let $x_c$ be a new variable. Since $\models \wedge s_1 \rightarrow \theta$ and $\models \wedge s_2 \rightarrow \neg\theta$, $\models \wedge s_1 \rightarrow \forall\{x_c \mid c \in C''\}\theta(\ldots c/x_c \ldots)$ and $\models \wedge s_2 \rightarrow \exists\{x_c \mid c \in C''\}\neg\theta(\ldots c/x_c \ldots)$. Hence if we let $\theta' = \forall\{x_c \mid c \in C''\}\theta(\ldots c/x_c \ldots)$, $\theta'$ is universal and satisfies (1) and (2) for $s_1$ and $s_2$, a contradiction.

(C6) Suppose $\{\exists v_i \sigma_i(v_i) \mid i \in I\} \subseteq s$. Then $\{\exists v_i \sigma_i(v_i) \mid i \in I\} \subseteq s_1$. Let $\{c_i \mid i \in I\}$ be a set of $|I|$ distinct constants of $C$ not appearing in $s$ and let $s_1' = s_1 \cup \{\sigma_i(c_i) \mid i \in I\}$, $s_2' = s_2$. Suppose $s' \notin S$, i.e. there is a universal $\theta$ which satisfies (1) and (2) for $s_1'$ and $s_2'$. Since no $c_i$ occurs in $s$, $\theta$ satisfies (1) and (2) for $s_1$ and $s_2$, a contradiction.

(C7b) Suppose $\{\sigma_i(t_i) \mid i \in I\} \cup \{c_i = t_i \mid i \in I\} \subseteq s$. Let

$I_1 = \{i \in I \mid \sigma_i(t_i) \in s_1,\ (c_i = t_i) \in s_1\}$;

$I_2 = \{i \in I \mid \sigma_i(t_i) \in s_1,\ (c_i = t_i) \in s_2,\ c_i \text{ occurs in } s_1\}$;

$I_3 = \{i \in I \mid \sigma_i(t_i) \in s_1,\ (c_i = t_i) \in s_2,\ c_i$ doesn't occur in $s_1\}$;

$I_4 = \{i \in I \mid \sigma_i(t_i) \in s_2,\ (c_i = t_i) \in s_2\}$;

$I_5 = \{i \in I \mid \sigma_i(t_i) \in s_2,\ (c_i = t_i) \in s_1,\ c_i$ occurs in $s_2\}$;

$I_6 = \{i \in I \mid \sigma_i(t_i) \in s_2,\ (c_i = t_i) \in s_1,\ c_i$ doesn't occur in $s_2\}$.

Let $s_1' = s_1 \cup \{\sigma_i(c_i) \mid i \in I_1 \cup I_2 \cup I_3\} \cup \{c_i = t_i \mid i \in I_2 \cup I_3\}$ and $s_2' = s_2$. If $s' \notin S$, there is a $\theta$ satisfying (1) and (2) for $s_1'$ and $s_2'$ so that all the constants of $M$ which appear in $\theta$ appear in $s_2$ and in $s_1$ or $\{c_i \mid i \in I_3\}$. Clearly $\models (\wedge s_1 \wedge \wedge\{c_i = t_i \mid i \in I_2 \cup I_3\}) \rightarrow \wedge s_1'$ so $\models \wedge s_1 \rightarrow (\wedge\{c_i = t_i \mid i \in I_2 \cup I_3\} \rightarrow \theta)$. For all $i \in I_3$, $c_i$ does not appear in $s_1$ or $\{c_i = t_i \mid i \in I_2\}$ so if $\theta'$ is like $\theta$ except that we replace each $c_j$, for $j \in I_3$, by $t_j$, $\models \wedge s_1 \rightarrow (\wedge\{c_i = t_i \mid i \in I_2\} \rightarrow \theta')$. Also since $i \in I_2$ implies $c_i = t_i \in s_2$, $\models \wedge s_2 \rightarrow \neg(\wedge\{c_i = t_i \mid i \in I_2\} \rightarrow \theta)$. But for $j \in I_3$, $c_j = t_j \in s_2$ and no $c_j$ occurs in $\{c_i = t_i \mid i \in I_2\}$, hence $\models \wedge s_2 \rightarrow \neg(\wedge\{c_i = t_i \mid i \in I_2\} \rightarrow \theta')$. If we let $\theta'' = \wedge\{c_i = t_i \mid i \in I_2\} \rightarrow \theta'$ then $\theta''$ is universal and satisfies (1) and (2) for $s_1$ and $s_2$, a contradiction. Similarly, if we let $s_1'' = s_1'$ and $s_2'' = s_2' \cup \{\sigma_i(c_i) \mid i \in I_4 \cup I_5 \cup I_6\} \cup \{c_i = t_i \mid i \in I_5 \cup I_6\}$ and suppose $\theta$ is universal and satisfies (1) and (2) for $s_1''$ and $s_2''$ then $\theta'' = \wedge\{c_i = t_i \mid i \in I_5\} \rightarrow \theta'$ (where $\theta'$ is like $\theta$ except we replace each occurrence of $c_j$, $j \in I_6$, by $t_j$) is universal and satisfies (1) and (2) for $s_1'$ and $s_2'$, another contradiction.

(C7c) Suppose $\{t_i \mid i \in I\}$ is a set of less than $\kappa$ constants of $M$. Let $\{c_i \mid i \in I\}$ be a set of $|I|$ distinct constants of $C$ not appearing in $s$ or $\{t_i \mid i \in I\}$. Let

$I_1 = \{i \in I \mid t_i$ appears in $s_1$ and $s_2$ or $t_i$ doesn't appear in $s_2\}$ and $I_2 = \{i \in I \mid t_i$ appears in $s_2\}$. If $s_k' = s_k \cup \{c_i = t_i \mid i \in I_k\}$ and $\theta$ is a universal sentence which satisfies (1) and (2) for $s_1'$

and $s_2'$, then every relation symbol of $L$ which appears in $\theta$ appears in $s_1$ and $s_2$ and every constant symbol of $M$ which appears in $\theta$ appears in $s_2$ and in either $s_1$, or $s_1$ and $s_2$, or not in $s_2$, and thus appears in $s_1$ and $s_2$. (If we had allowed function symbols in $L$ we could not have used this argument.) In particular, for every $i \in I$, $c_i$ does not appear in $\theta$, so that $c_i$ can only occur at $c_i = t_i$ in $\wedge s_1' \longrightarrow \theta$ and $\wedge s_2' \longrightarrow \neg\theta$. Hence $\models \wedge s_1 \longrightarrow \theta$ and $\models \wedge s_2 \longrightarrow \neg\theta$ which contradicts $s \in S$.

Since $\{\varphi, \neg\psi\}$ has no model it is not an element of $S$. Hence there is a universal sentence $\theta$ of $M_{2^{*\kappa}\kappa}$ which satisfies (1) and (2) for $\{\varphi\}$ and $\{\neg\psi\}$. But no $c \in C$ appears in either $\varphi$ or $\neg\psi$ so no $c \in C$ appears in $\theta$. Thus $\theta$ is a sentence of $L_{2^{*\kappa}\kappa}$ satisfying the desired conditions. ⊣

The Malitz interpolation theorem is clearly equivalent to the theorem which states that if $\varphi$ is existential the interpolating sentence is existential. The proof of the Craig-Lyndon interpolation theorem showed that, if $L$ has no function or constant symbols and if $\varphi$ is universal then the interpolating sentence is universal, and therefore also if $\psi$ is existential the interpolating sentence is existential.

3.5 <u>Preservation Under Submodels and Extensions</u>. We say a sentence $\varphi$ is *preserved under submodels (extensions) relative to* $\sigma$ iff whenever $\mathfrak{A}$ is a model of $\varphi \wedge \sigma$, every submodel (extension) of $\mathfrak{A}$ which is a model of $\sigma$ is a model of $\varphi$.

The main result of this section is that a sentence is preserved under submodels (relative to $\sigma$) iff there is a universal sentence, of an appropriate language, to which it is equivalent (in every model of $\sigma$). To show this for languages which allow uncountable conjunctions we need a stronger version of the Downward Löwenheim-Skolem Theorem which we just state here.

DOWNWARD LÖWENHEIM-SKOLEM-TARSKI THEOREM. *Suppose* $\kappa$ *is any infinite cardinal and* $\varphi$ *is a sentence of* $L_{\kappa^+\omega}$. *If* $\mathfrak{A}$ *is a model of* $\varphi$ *then there is a submodel* $\mathfrak{B}$ *of* $\mathfrak{A}$ *such that* $|B| \leq \kappa$ *and* $\mathfrak{B}$ *is a model of* $\varphi$.

COROLLARY. *Suppose* $\sigma$ *and* $\varphi$ *are sentences of* $L_{\kappa\omega}$, $\kappa > \omega$ *and* $\kappa$ *regular or* $cf(\kappa) = \omega$. $\varphi$ *is preserved under submodels relative to* $\sigma$ *iff there is a universal sentence* $\theta$ *of* $L_{2^{*}\kappa_\kappa}$ ($L_{\omega_1\omega}$ *if* $\kappa = \omega_1$) *such that* $\models \sigma \longrightarrow (\varphi \longleftrightarrow \theta)$.

PROOF. Assume $L$ has no function symbols by replacing function symbols by relation symbols in the usual way. If $\kappa$ is regular and $\kappa > \omega$, there is a $\lambda < \kappa$ such that $\sigma$ and $\varphi$ are sentences of $L_{\lambda^+\omega}$. Let $D$ be a set of constants which do not occur in $\sigma$ or $\varphi$, of cardinality $\lambda$ if $\kappa$ is regular and $cf(\kappa) > \omega$, or of cardinality $\kappa$ if $cf(\kappa) = \omega$. For each formula $\psi$ of $L_{\lambda^+\omega}$ ($L_{\kappa\omega}$ if $cf(\kappa) = \omega$) define the quantifier free formula $\psi^D$ of $L_{\kappa\omega}$ as follows:

$$\psi^D = \psi, \text{ for } \psi \text{ atomic}$$

$$(\neg\psi)^D = \neg(\psi^D)$$

$$(\wedge\Psi)^D = \wedge\{\psi^D \mid \psi \in \Psi\}$$

$$(\vee\Psi)^D = \vee\{\psi^D \mid \psi \in \Psi\}$$

$$(\forall x \psi(x))^D = \wedge\{\psi^D(d) \mid d \in D\}$$

$$(\exists x \psi(x))^D = \vee\{\psi^D(d) \mid d \in D\}.$$

If $\varphi$ is preserved under submodels relative to $\sigma$, $\models (\sigma \wedge \varphi) \longrightarrow (\sigma^D \longrightarrow \varphi^D)$. Since $\sigma^D \longrightarrow \varphi^D$ is quantifier free it is universal and, since we are assuming $L$ has no function symbols, we can apply Theorem 3.4 to the above sentences of $L_{\kappa\omega}$. I.e., there is a universal sentence $\theta$ of $L_{2^{*\kappa}\kappa}$ ($L_{\omega_1\omega}$ if $\kappa = \omega_1$) such that $\models (\sigma \wedge \varphi) \longrightarrow \theta$ and $\models \theta \longrightarrow (\sigma^D \longrightarrow \varphi^D)$. Suppose $\sigma \longrightarrow (\theta \longrightarrow \varphi)$ is not valid, i.e. there is a model $\mathfrak{A}$ of $\sigma \wedge \theta \wedge \neg\varphi$. By the Downward Löwenheim-Skolem-Tarski Theorem there is a submodel $\mathfrak{B}$ of $\mathfrak{A}$ such that $\mathfrak{B}$ is a model of $\sigma \wedge \neg\varphi$ and $|B| \leq \lambda$ ($|B| \leq \kappa$ if $cf(\kappa) = \omega$). Clearly $B$ can be enumerated by $D$ so $\mathfrak{B}$ is a model of $\sigma^D \wedge \neg\varphi^D$ and hence $\mathfrak{B}$ is a model of $\neg\theta$. But $\neg\theta$ is an existential sentence and $\mathfrak{B}$ is a submodel of $\mathfrak{A}$ so that $\mathfrak{A}$ is a model of $\neg\theta$, a contradiction. Hence $\models \sigma \longrightarrow (\varphi \longleftrightarrow \theta)$ if $\varphi$ is preserved under submodels relative to $\sigma$.

Suppose now there is such a $\theta$ and $\mathfrak{A}$ is a model of $\sigma \wedge \varphi$, $\mathfrak{B}$ is a model of $\sigma$ and $\mathfrak{B}$ is a submodel of $\mathfrak{A}$. Then $\mathfrak{A}$ is a model of $\theta$ and since $\theta$ is universal $\mathfrak{B}$ is a model of $\theta$, so $\mathfrak{B}$ is a model of $\varphi$. ⊣

Note that if $\kappa$ is strongly inaccessible the interpolating sentences are in $L_{\kappa\kappa}$ and the interpolating sentence of the last theorem can be found to be prenex universal. In this case the universal sentence of the above corollary can also be found to be a prenex universal sentence of $L_{\kappa\kappa}$.

If $\kappa$ is a strong limit cardinal of cofinality $\omega$, and $\varphi$ and $\sigma$ are sentences of $L_{\kappa^+\omega}$ then $\theta$ can be found in $L_{\kappa^+\kappa}$.

Since $\varphi$ is preserved under extensions relative to $\sigma$ iff $\neg\varphi$ is preserved under submodels relative to $\sigma$, it is clear that $\varphi$ is preserved under extensions relative to $\sigma$ iff there is an existential sentence $\theta$ of $L_{2^{*\kappa}\kappa}$ such that $\models \sigma \longrightarrow (\varphi \longleftrightarrow \theta)$.

4. <u>$(\kappa,\omega)$ Consistency Properties</u>. In this section we introduce a

modification of the $\kappa$ consistency properties of the previous sections. We distinguish a countable set of constants and adapt the closure conditions to allow treatment of only finitely many of these constants at a time. To do this we let D be a countable set of constants none of which appear in L or C and let N be the language L together with C and D.

4.1 <u>Definition</u>. Suppose $\kappa$ is regular, $\kappa \neq \omega_1$ and S is a set of sets of sentences of $N_{\kappa\omega}$. S is a $(\kappa,\omega)$ *consistency property* iff S satisfies Definition 1.1 (*ii*) with M replaced by N and only finitely many $d \in D$ added to $s \in S$ for (C2) - (C7).

If $cf(\kappa) = \omega$ and $\kappa$ is singular, in order to define a $(\kappa,\omega)$ consistency property we make the above changes to Definition 1.1 (*iii*).

By slightly modifying the proof of Theorem 1.2 we can see that:

MODEL EXISTENCE THEOREM. *If* S *is a* $(\kappa,\omega)$ *consistency property and* $s \in S$ *with every sentence in* s *having only finitely many* $d \in D$ *appearing in it, then* s *has a model.*

4.2 THEOREM. *Suppose* $\kappa$ *and* $2^{*\kappa}$ *are regular* $(\kappa \neq \omega_1)$, *e.g.* $\kappa$ *is strongly inaccessible or* $\kappa$ *is a successor cardinal. Suppose* $\theta$ *is a sentence of* $L_{\kappa\omega}$, $\underline{U}$ *and* $<$ *are unary and binary relation symbols of* L *and for all* $\alpha < 2^{*\kappa}$, $\theta$ *has a model* $\mathfrak{A} = \langle A,\underline{U},<,\ldots\rangle$ *such that* $<$ *is a transitive linear ordering of* $\underline{U}$ *and* $\langle\alpha,<\rangle \subseteq \langle\underline{U},<\rangle$. *Then* $\theta$ *has a model* $\mathfrak{B} = \langle B,\underline{V},<,\ldots\rangle$ *such that* $<$ *linearly orders* $\underline{V}$ *and* $\langle\underline{V},<\rangle$ *is not well ordered.*

PROOF. Write D as $\{d_r \mid r \in Q\}$, where $Q$ is the set of rational numbers. Let M be the class of all models which interpret L such that if $\mathfrak{A} = \langle A,\underline{U},\ ,\ldots\rangle \in M$ then $\underline{U}'$, the set of ordinals in $\underline{U}$, is an ordinal which is less than $2^{*\kappa}$ and $<$ linearly orders $\underline{U}$.

Let S be the set of all sets s such that $s = s_0 \cup \{d_q < d_r \mid q < r,\ q,r \in \mathcal{Q}\} \cup \{\underline{U}d_q \mid q \in \mathcal{Q}\}$ where $s_0$ is a set of sentences of $N_{\kappa\omega}$ of cardinality less than $\kappa$ in which only finitely many $d \in D$ appears and such that if $d_{r_1},\ldots,d_{r_n}$ are the only constants from D appearing in $s_0$ and $r_1 < \ldots < r_n$, then for all $\alpha < 2^{*\kappa}$ there is an $\mathfrak{A} \in M$, there are $b_1,\ldots,b_n \in \underline{U}'$, there is an assignment $s$ of elements of A to constants of C such that

$$\mathfrak{A},b_1,\ldots,b_n,s \models \wedge s_0, \quad \langle\alpha,<\rangle \subseteq \langle\underline{U}',<\rangle, \quad \mathfrak{A} \models \theta,$$

and

$$(\alpha \le b_1),(b_1 + \alpha \le b_2),\ldots,(b_{n-1} + \alpha \le b_n).$$

We claim S is a $(\kappa,\omega)$ consistency property and suppose $s \in S$ with $s = s_0 \cup \{d_q < d_r \mid q < r,\ q,r \in \mathcal{Q}\} \cup \{\underline{U}d_q \mid q \in \mathcal{Q}\}$.

(C1). Suppose $\{\varphi,\neg\varphi\} \subseteq s$. Since $s \in S$ implies $s_0$ has a model, $\{\varphi,\neg\varphi\}$ is not a subset of $s_0$. Hence $\varphi$ is $d_{r_1} < d_{r_2}$ for some $r_1 < r_2$, or $\varphi$ is $\underline{U}d_q$ for some $q \in \mathcal{Q}$. In the first case, $\neg(d_{r_1} < d_{r_2}) \in s_0$ and we can find $\mathfrak{A},b_1,b_2$ such that $b_1 < b_2$ and $\mathfrak{A},b_1,b_2 \models \neg(d_{r_1} < d_{r_2})$. In the second case, $\neg(\underline{U}d_q) \in s_0$ and we can find an $\mathfrak{A}$, b, such that $b \in \underline{U}'$ and $\mathfrak{A},b \models \neg\underline{U}d_q$. Both of these cases are contradictory.

In (C2)-(C6) and (C7a) all the formulas we consider are not atomic formulas of the form $d_q < d_r$ or $\underline{U}d_q$, so we need only consider subsets of $s_0$. But then it is clear that all but cases (C5), (C7b) and (C7c) are trivial.

(C5) Suppose $\{\vee\Phi_i \mid i \in I\} \subseteq s_0$. We know that for all $\alpha < 2^{*\kappa}$ there is an $\mathfrak{A}_\alpha \in M$, there are $b_{1,\alpha},\ldots,b_{n,\alpha} \in \underline{U}'_\alpha$, there is an assignment $s_\alpha$ such that $\langle\alpha,<\rangle \subseteq \langle\underline{U}'_\alpha,<\rangle$, $\mathfrak{A}_\alpha \models \theta$, $(\alpha \le b_{1,\alpha}),(b_{1,\alpha}+\alpha \le b_{2,\alpha}),\ldots,(b_{n-1,\alpha}+\alpha \le b_{n,\alpha})$ and $\mathfrak{A}_\alpha, b_{1,\alpha},\ldots,b_{n,\alpha},s_\alpha \models \wedge s_0$. Since the distributive law is valid, for all $\alpha < 2^{*\kappa}$ there is an $f_\alpha \in X\{\Phi_i \mid i \in I\}$ such that $\mathfrak{A}_\alpha, b_{1,\alpha},\ldots,b_{n,\alpha},s_\alpha \models \wedge s_0 \wedge \wedge\{f_\alpha(i) \mid i \in I\}$. But since $|X\{\Phi_i \mid i\in I\}| < 2^{*\kappa}$ and $2^{*\kappa}$ is regular, there is an $f \in X\{\Phi_i \mid i \in I\}$ such that for arbitrarily large $\alpha < 2^{*\kappa}$, $\mathfrak{A}_\alpha, b_{1,\alpha},\ldots,b_{n,\alpha},s_\alpha \models \wedge s_0 \wedge \wedge\{f(i) \mid i\in I\}$. For such an $f$ let $s'_0 = s_0 \cup \{f(i) \mid i \in I\}$ and suppose $\beta < 2^{*\kappa}$. Let $\alpha(\beta)$ be the least $\alpha \ge \beta$ for which the above holds and let $\mathfrak{A}'_\beta = \mathfrak{A}_{\alpha(\beta)}$, $b'_{m,\beta} = b_{m,\alpha(\beta)}$, and $s'_\beta = s_{\alpha(\beta)}$. Then $(\beta \le \alpha(\beta) \le b_{1,\alpha(\beta)} = b'_{1,\beta}),\ldots,(b'_{n-1,\beta}+\beta \le b_{n-1,\alpha(\beta)}+\alpha(\beta) \le b_{n,\alpha(\beta)} = b'_{n,\beta})$ and $\mathfrak{A}'_\beta, b'_{1,\beta},\ldots,b'_{n,\beta},s'_\beta \models \wedge s'_0$.

(C7b) Suppose $\{\psi_i(t_i) \mid i\in I\} \cup \{c_i = t_i \mid i \in I\} \subseteq s$, and only finitely many $d \in D$ appear in $\{\psi_i(t_i) \mid i \in I\}$. Let $I_1 = \{i\in I \mid \psi_i(t_i) \in s_0\}$, $I_2 = \{i \in I \mid \psi_i(t_i) \in \{d_q < d_r \mid q < r\}\}$, and $I_3 = \{i\in I \mid \psi_i(t_i) \in \{\underline{U}d_q \mid q \in Q\}\}$. Then $I = I_1 \cup I_2 \cup I_3$ and if $s'_0 = s_0 \cup \{\psi_i(c_i) \mid i \in I_1 \cup I_3\}$, $\mathfrak{A}, b_1,\ldots,b_n, s \models \wedge s_0$ implies $\mathfrak{A}, b_1,\ldots,b_n,s \models \wedge s'_0$. Since only finitely many $d \in D$ appear in $\{\psi_i(t_i) \mid i \in I_2\}$ and every $\psi_i(t_i)$ for $i \in I_2$ is of the form $d_q < d_r$, we know $I_2$ is finite and hence we may assume it has only one element, say $i$. But then $\psi_i(t_i)$ is $t_i < d_r$ or $d_r < t_i$ and $t_i$ is $d_{r_k}$ for some $k \in \{1,\ldots,n\}$. If $r \in \{r_1,\ldots,r_n\}$, $\mathfrak{A}, b_1,\ldots,b_n, s \models \wedge s_0$, and $b_1 < \ldots < b_n$ then $\mathfrak{A}, b_1,\ldots,b_n \models \psi_i(t_i)$ and hence $\mathfrak{A}, b_1,\ldots,b_n, s \models \psi_i(c_i)$. Suppose $r \notin \{r_1,\ldots,r_n\}$, then for some $m \in \{0,1,\ldots,n\}$, $r_m < r < r_{m+1}$. Hence for all $\alpha < 2^{*\kappa}$, if $\mathfrak{A}_\alpha, b_{1,\alpha},\ldots,b_{n,\alpha},s_\alpha$ satisfies the conditions for $s'_0$ and $\alpha$, we know $\mathfrak{A}_{\alpha+\alpha}, b_{1,\alpha+\alpha},\ldots,b_{m,\alpha+\alpha}, b, b_{m+1,\alpha+\alpha},\ldots,b_{n,\alpha+\alpha},s_{\alpha+\alpha}$ satisfies the conditions for $s'_0 \cup \{\psi_i(c_i)\}$ and $\alpha$, where $b$ is $b_{m,\alpha+\alpha}+\alpha$. So we can find for each $\alpha < 2^{*\kappa}$ a model which satisfies the conditions for $\alpha$ and

$s_0'' = s_0' \cup \{\psi_i(c_i) \mid i \in I_2\} = s_0 \cup \{\psi(c_i) \mid i \in I\}$.

(C7c) Suppose $\{t_i \mid i \in I\}$ is a set of less than $\kappa$ basic terms of N w.r.t. C such that only finitely many $d \in D$ appear in it. Let $I_1 = \{i \in I \mid t_i \notin D\}$ and $I_2 = I - I_1$. Let $\{c_i \mid i \in I\}$ be a set of $|I|$ distinct constants none of which appear in $s_0$ or $\{t_i \mid i \in I\}$. Then if $\mathfrak{A}, b_1, \ldots, b_n, s \models \wedge s_0$, $\mathfrak{A}, b_1, \ldots, b_n, s' \models \wedge s_0'$ where $s_0' = s_0 \cup \{c_i = t_i \mid i \in I_1\}$ and $s' = (s - \{(c_i, s(c_i)) \mid i \in I_1\}) \cup \{(c_i, s(t_i)) \mid i \in I_1\}$. Since for all $i \in I_2$, $t_i$ is an element of D, we know $I_2$ is finite and hence we can find interpretations for the $c_i$, $i \in I_2$, one at a time. If $t_i$ appears in $s_0$, say $t_i$ is $d_{r_{k(i)}}$, then for a given $s$ let $s'' = (s' - \{(c_i, s(c_i))\}) \cup \{(c_i, b_{k(i)})\}$. Suppose $t_i$ doesn't appear in $s_0$. Then $t_i$ is $d_r$ for some $r \notin \{r_1, \ldots, r_n\}$ and as in C7b we can position $r$ among the $r_m$ and interpret $c_i$ by an appropriate b.

This completes the proof that S is a $(\kappa, \omega)$ consistency property. It remains only to show that it is the one we want.

Suppose $\psi$ is a sentence of $L_{\omega\omega}$ which says $<$ is a transitive linear ordering of $\underline{U}$. We claim

$$\{\theta, \psi\} \cup \{d_q < d_r \mid q < r,\ q \in Q,\ r \in Q\} \cup \{\underline{U} d_q \mid q \in Q\} \in S.$$

Suppose $\alpha < 2^{*\kappa}$, by the hypothesis of the theorem there is a model $\mathfrak{A}$ of $\{\theta, \psi\}$ such that $\langle \alpha, < \rangle \subseteq \langle \underline{U}, < \rangle$. For each $\xi \in \alpha$ add a new constant $e_\xi$ to L and call the new language $L^\alpha$. Then if $\varphi_\alpha$ is the sentence of $L^\alpha_{2^{*\kappa}\kappa}$,

$$\wedge\{e_\xi < e_{\xi'} \mid \xi < \xi' < \alpha\} \wedge \wedge\{\underline{U} e_\xi \mid \xi < \alpha\} \wedge \theta \wedge \psi,$$

$\mathfrak{A}, s \models \varphi_\alpha$ for the assignment $\{(e_\xi, \xi) \mid \xi \in \alpha\}$. Since $2^{*\kappa}$ is regular, $\varphi_\alpha$ has a model $\mathfrak{A}_\alpha$ of cardinality less than $2^{*\kappa}$. But then the set of ordinals in $\underline{U}_\alpha$ is an ordinal less than $2^{*\kappa}$, so $\mathfrak{A}_\alpha \in M$.

Since $\mathfrak{A}_\alpha \models \varphi_\alpha$, we have $\langle\alpha,<\rangle \subseteq \langle \underline{U}'_\alpha,<\rangle$ and $\mathfrak{A}_\alpha \models \theta \wedge \psi$. Thus our claim is proved and therefore $\{\theta,\psi\} \cup \{d_q < d_r \mid q < r,\ q\in Q,\ r\in Q\} \cup \{\underline{U}d_q \mid q\in Q\}$ has a model $\mathfrak{B} = \langle B,\underline{V},<,\ldots\rangle$ such that $<$ linearly orders $\underline{V}$ and $\langle\underline{V},<\rangle$ contains a copy of the rationals. ⊣

COROLLARY (Lopez-Escobar [21]). *Well ordering cannot be characterized by any sentence of* $L_{\infty\omega}$.

PROOF. Suppose $\varphi$ is a sentence of $L_{\infty\omega}$ such that for every $\mathfrak{A} = \langle A,<,\ldots\rangle$, $\mathfrak{A} \models \varphi$ iff $<$ well orders $A$. Choose a unary predicate $\underline{U}$ which does not appear in $\varphi$ and let $\theta$ be $\varphi \wedge \forall x \underline{U}(x)$. Then $\theta$ is a sentence of $L_{\lambda^+\omega}$ for some $\lambda$. But for each $\alpha < (2^\lambda)^+$ the model $\mathfrak{A}_\alpha = \langle A_\alpha, \underline{U}_\alpha, <,\ldots\rangle$ satisfies the hypothesis of Theorem 4.2 for $\theta$ and $A_\alpha = \underline{U}_\alpha = \alpha$. Hence there is a model $\mathfrak{B} = \langle B,\underline{V},<,\ldots\rangle$ of $\varphi$ such that $B = \underline{V}$ and $B$ is not well ordered, a contradiction. ⊣

4.3. Hanf Numbers. The *Hanf number* of a language is the least cardinal $\lambda$ such that any sentence of the language which has a model of cardinality at least $\lambda$, has arbitrarily large models.

Barwise and Kunen [3] have shown that, for any $\kappa$, the Hanf number of $L_{\kappa\omega} \cap H_\kappa$ is $\beth_{h(\kappa)}$, where $h(\kappa)$ is the least ordinal which will work as a bound in Theorem 4.2 for the language $L_{\kappa\omega}$. Hence we can find bounds on Hanf numbers using Theorem 4.2 and the following variations on that theorem.

(a) Theorem 4.2 holds for $\kappa$ regular and $2^{*\kappa}$ singular.

(b) If $cf(\kappa) = \omega$ and $\kappa > \omega$ then Theorem 4.2 holds with a bound of $(2^{*\kappa})^+$. It holds with this same bound for sentences of $L_{\kappa^+\omega}$.

(c) If $\kappa = \omega_1$ then Theorem 4.2 holds with a bound of $\omega_1$.

(a) holds since if $2^{*\kappa}$ is singular and $\kappa$ is regular, any sentence of $L_{\kappa\omega}$ is a sentence of $L_{\lambda^+\omega}$ for some $\lambda$ such that $(2^\lambda)^+ < 2^{*\kappa}$ and Theorem 4.2 holds for the language $L_{\lambda^+\omega}$. Obvious variations of the consistency properties defined in the proof of Theorem 4.2 can be used to show (b) and (c). Thus we have:

If $\kappa$ is regular and $\kappa \neq \omega_1$, the Hanf number of $L_{\kappa\omega}$ is no larger than $\beth_{2^{*\kappa}}$.

It is easy to see, using an example like Example 18 in Keisler [18], that for any $\kappa$ the Hanf number of $L_{\kappa\omega}$ is greater than or equal to $\beth_\kappa$. So if $\kappa$ is strongly inaccessible or $\kappa = \omega_1$, the Hanf number of $L_{\kappa\omega}$ is $\beth_\kappa$.

For $cf(\kappa) \neq \omega$, Theorem 4.2 says the Hanf number of $L_{\kappa^+\omega}$ is no larger than $\beth_{(2^\kappa)^+}$. It is in fact known that the Hanf number of $L_{\kappa^+\omega}$ is strictly less than $\beth_{(2^\kappa)^+}$ (see [5]). If however, $\kappa$ is a strong limit cardinal of cofinality $\omega$, the Hanf number of $L_{\kappa^+\omega}$ is $\beth_{\kappa^+}$ (Helling [12]) because the bound of Theorem 4.2 can be reduced to $\kappa^+$. The question then arises, for what other $\kappa$ will the bound $\kappa^+$ work in Theorem 4.2? A partial answer is that it will not if $\kappa^{\aleph_0} = \kappa$. To see this look at a language $L_{\kappa^+\omega}$ where L contains $\underline{U}$, $<$, and a set of constants B of cardinality $\kappa$. Let $\theta$ be the conjunction of: "$<$ is a linear ordering of $\underline{U}$", $\forall v \bigvee\{v = b \mid b \in B\}$, and $\sim\bigwedge\{s(x_{n+1}) < s(x_n) \mid n \in \omega\}$ for all assignments $s$ of elements of B to the variables $\{x_n \mid n \in \omega\}$. Since the cardinality of the set of assignments of elements of B to the variables $\{x_n \mid n \in \omega\}$ is $\kappa^{\aleph_0}$ and $\kappa^{\aleph_0} = \kappa$, we have $\theta$ is a sentence of $L_{\kappa^+\omega}$. But clearly in any model $\mathfrak{A}$ of $\theta$, $<$ well orders $\underline{U}$ and this well ordering can have any length less than $\kappa^+$. This in fact shows that assuming $\kappa$ is a strong limit cardinal, the bound $\kappa^+$ works iff $cf(\kappa) = \omega$.

# CHAPTER II

## FRAGMENTS OF $L_{\infty\omega}$

In the previous chapter our only method of classifying formulas of a language was by the cardinality of the sets of formulas over which we took conjunctions and disjunctions. At this point we introduce the notion of a fragment of a language $L_{\infty\omega}$, i.e. a set of formulas closed under subformulas and finitary logical operations, in order to identify certain sublanguages of $L_{\infty\omega}$ that satisfy compactness theorems. We first formalize within set theory, in the usual manner, the notion of a finite-quantifier language.

1. Preliminaries. As *symbols* of our language we have for some sets x and some natural numbers m (which sets and which natural numbers depend on the language)

(a) variables $(0,x)$ written $v_x$,

(b) constants $(1,x)$ and $(2,x)$ denoted by c, d, etc.,

(c) m-ary relation symbols $(3,x,m)$ denoted by $\underline{R}_x$, $\underline{R}$, $\underline{S}$, etc.,

(d) m-ary function symbols $(4,x,m)$ denoted by $\underline{f}_x$, $\underline{f}$, $\underline{g}$, etc..

The class of *terms* of L is the least class such that all constants and variables are terms and if $\underline{f}$ is an m-ary function symbol, $t_1,\ldots,t_m$ are terms, then $(\underline{f},t_1,\ldots,t_m)$, written $\underline{f}(t_1,\ldots,t_m)$, is a term.

*Atomic formulas* are of the form $(\underline{R},t_1,\ldots,t_m)$, written $\underline{R}(t_1,\ldots,t_m)$, where $\underline{R}$ is an m-ary relation symbol and $t_1,\ldots,t_m$ are terms.

The class of *formulas* of L is the last class including all the atomic formulas such that

(a) if $\varphi$ is a formula then $(5,\varphi)$, written $\neg\varphi$, is a formula,

(b) if $\varphi$ is a formula and $v$ is a variable $(6,v,\varphi)$, written $\forall v\varphi$, and $(7,v,\varphi)$, written $\exists v\varphi$, are formulas,

(c) if $\Phi$ is a set of formulas, $(8,\Phi)$ and $(9,\Phi)$, written $\wedge\Phi$ and $\vee\Phi$ are formulas.

Other notation that we will use includes:

$\langle a_i \mid i\in I\rangle$ for the function $\{(i,a) \mid i\in I\}$,

$(u)_{n-1}$ for the $n^{\underline{th}}$ entry in an ordered n-tuple $u$.

Although we are interested mainly in the semantic properties of languages we must also be assured of their nice syntactic properties. Thus we introduce the notion of primitive recursive set functions [13] in addition to the hierarchy of set theoretic predicates [19].

<u>Definition</u>. A set function is *primitive recursive (Prim)* if it can be obtained from the initial functions by substitutions and recursion as follows:

Initial functions

(1) $F(v_1,\ldots,v_n) = v_m$ for each $m \le n \in \omega$

(2) $F(v) = 0$

(3) $F(v,w) = \{v,w\}$

(4) $F(v,w) = v - w$

(5) $F(v) = \cup v$

Substitution

$F(v_1,\ldots,v_n) = G(H_1(v_1,\ldots,v_n),\ldots,H_p(v_1,\ldots,v_n))$ for $n,p \in \omega$, $G,H_1,\ldots,H_p$ Prim

Recursion

$F(w,v_1,\ldots,v_n) = G(F\upharpoonright w,w,v_1,\ldots,v_n)$ for $n \in \omega$, $G$ Prim, and

$F\upharpoonright w = \{(u,F(u,v_1,\ldots,v_n)) \mid u\in w\}$.

If $\mathcal{F}$ is a one-place set function, a set function is *primitive recursive in* $\mathcal{F}$ if it satisfies the above definition with the additional initial function $F(v) = \mathcal{F}(v)$ and Prim replaced by Prim $\mathcal{F}$.

Definition. A finite formula involving $=$ and the two place relation symbol $\in$ is called $\Delta_0$ if all of its quantifiers are restricted, i.e. are of the form $(\forall v)(v\in w\rightarrow\ldots)$ or $(\exists v)(v\in w\wedge\ldots)$. A formula is $\Sigma_1$ $(\Pi_1)$ if it is of the form $\exists v\psi$ $(\forall v\psi)$ where $\psi$ is a $\Delta_0$ formula. A formula is called generalized $\Sigma_1$ if it is built up from $\Delta_0$ formulas by finitary conjunction and disjunction and by existential and bounded quantification.

A relation $R$ on a set $A$ is called $\underset{\sim}{\Delta}_0(\underset{\sim}{\Sigma}_1,\underset{\sim}{\Pi}_1)$ definable on $A$ iff there is a $\Delta_0(\Sigma_1,\Pi_1)$ formula $\psi(v_1,\ldots,v_n,w_1,\ldots,w_m)$ and $b_1,\ldots,b_m \in A$ such that $R(a_1,\ldots,a_n)$ iff $\langle A,\in\rangle \models \psi[a_1,\ldots,a_n,b_1,\ldots,b_m]$. We call a relation on $A$ $\underset{\sim}{\Delta}$ definable on $A$ iff it is both $\underset{\sim}{\Sigma}_1$ definable on $A$ and $\underset{\sim}{\Pi}_1$ definable on $A$.

For a finite formula which involves an additional relation symbol $\underline{R}$ we define $\Delta_0$ in $\underline{R}$ and the subsequent notions in the obvious manner.

We call a relation primitive recursive (in $\mathcal{F}$) iff its characteristic function is Prim (Prim $\mathcal{F}$). The primitive recursive set functions are closed under definition by cases while the primitive recursive relations are closed under Boolean operations and bounded quantifications. Thus any relation which is $\underset{\sim}{\Delta}_0$ definable on a Prim closed set is primitive recursive.

If we now return to our formal language we can easily see that the following functions and relations are primitive recursive if the sets of relation, constant, and function symbols are $\underset{\sim}{\Delta}_0$ definable:

AtFmla(y) - y is an atomic formula

Fmla(y) - y is a formula

Neg(y) - y is a negation

Conj(y) - y is a conjunction

Exist(y) - y is an existential quantification

$\lambda y(FV(y))$ - the set of free variables appearing in y if Fmla(y)

$\lambda y,f(SF(y,f))$ - the result of substituting f(v) for free occurrences of v in y if Fmla(y), Fcn(f), $v \in Dm(f) \cap FV(y)$, and f(v) a term.

Write $SF(\varphi,v,x)$ for $SF(\varphi,\{(v,x)\})$.

For any transitive A which is closed under all primitive recursive set functions the language $L_A = L_{\infty\omega} \cap A$ is closed under subformulas and finitary logical operations, i.e. is a fragment of $L_{\infty\omega}$.

In order to consider the nice semantic properties of $L_A$ we consider the notion of an admissible set due to Platek [24]. These particular transitive Prim closed sets come into the study of infinitary languages $L_{\infty\omega}$ in connection with proofs in ordinary predicate logic extended to $L_{\infty\omega}$. If A is admissible, a sentence of $L_A$ which is provable is provable in A, and if $\Phi$ is a $\underset{\sim}{\Sigma}_1$ on A set of sentences, a sentence which is deducible from $\Phi$ is deducible from $\Phi$ in A. However only in the case A is countable is this logic complete. If $\mathcal{P}$ is the power set operation, uncountable admissible and $\mathcal{P}$-admissible sets play a similar role in connection with complete logics which include distributive laws and a rule of disjunctive generalization.

The usual definition of admissibility is equivalent to the following.

Definition. A set $A$ is *admissible* iff it is transitive, Prim closed and satisfies the $\underset{\sim}{\Delta}_0$ collection scheme, i.e. for any $n+2$ place $\Delta_0$ formula $\psi$ if $a_1,\ldots,a_n,x \in A$ and

$$(\forall u\in x)(\exists v\in A)(\langle A,\in\rangle \models \psi[u,v,a_1,\ldots,a_n])$$

then

$$(\exists w\in A)(\forall u\in x)(\exists v\in w)(\langle A,\in\rangle \models \psi[u,v,a_1,\ldots,a_n]).$$

If $\mathcal{P}$ is the power set operation and $\mathfrak{G}_{\mathcal{P}}$ is its graph, we call a set *$\mathcal{P}$-admissible* iff it is transitive, Prim $\mathcal{P}$ closed and satisfies the $\underset{\sim}{\Delta}_0$ in $\mathfrak{G}_P$ collection scheme, i.e. for any $n+2$ place $\Delta_0$ in $\mathfrak{G}_{\mathcal{P}}$ formula $\psi$ if $a_1,\ldots,a_n,x \in A$ and

$$(\forall u\in x)(\exists v\in A)(\langle A,\in,\mathfrak{G}_{\mathcal{P}}\rangle \models \psi[u,v,a_1,\ldots,a_n])$$

then

$$(\exists w\in A)(\forall u\in x)(\exists v\in w)(\langle A,\in,\mathfrak{G}_{\mathcal{P}}\rangle \models \psi[u,v,a_1,\ldots,a_n]).$$

If in addition $A = \bigcup\{A_n \mid n\in\omega\}$ and for all $n \in \omega$, $A_n \in A$, we call $A$ a *cofinality* $\omega$ *$\mathcal{P}$-admissible* set. We can clearly assume that if $A$ is a cf $\omega$ $\mathcal{P}$-admissible set then each $A_n$ is transitive and a subset of $A_{n+1}$.

2. <u>A Consistency Properties</u>. For the remainder of this article we assume $A = \bigcup\{A_n \mid n\in\omega\}$ with $A_n \subseteq A_{n+1}$ and the new constants $C$ are enumerated by $A$, i.e. we assume no constant $(2,x)$ appears in $L$ and let $C_n = \{(2,a) \mid a\in A_n\}$ and $C = \bigcup\{C_n \mid n\in\omega\}$. For each $a \in A$, we write $c_a$ for the constant $(2,a)$. We now define a consistency property for those fragments $L_A$ for which the symbols are $\underset{\sim}{\Delta}_0$ definable. This assumption on the symbols will be in force

throughout, as will the assumption $|A_n| < |A_{n+1}| < |A|$.

2.1 Definition. Suppose $L_A$ is a fragment of $L_{\infty\omega}$. S is an A *consistency property* iff for each $s \in S$, s is a set of sentence of $M_A$, the constants of C which appear in s all come from one $C_n$, and all of the following hold:

(C1) If $\varphi$ is a sentence of $M_A$, then either $\varphi \notin s$ or $\neg\varphi \notin s$.

(C2) If $\{\neg\varphi \mid \varphi \in \Phi\} \subseteq s \cap A_n$ for some $n \in \omega$, then $s \cup \{\varphi\neg \mid \varphi \in \Phi\} \in S$.

(C3) If $\{\wedge\Phi_i \mid i \in I\} \subseteq s \cap A_n$ for some $n \in \omega$, then $s \cup \bigcup\{\Phi_i \mid i \in I\} \in S$.

(C4) If $\{\forall v_i \varphi_i(v_i) \mid i \in I\} \subseteq s \cap A_n$ for some $n \in \omega$ and $B \subseteq A_m$ for some $m \in \omega$, then $s \cup \{\varphi_i(c_b) \mid i \in I, b \in B\} \in S$.

(C5) If $\{\vee\Phi_i \mid i \in I\} \subseteq s \cap A_n$ for some $n \in \omega$, then there is an $f \in \times\{\Phi_i \mid i \in I\}$ such that $s \cup \{f(i) \mid i \in I\} \in S$.

(C6) If $\{\exists v_i \varphi_i(v_i) \mid i \in I\} \subseteq s \cap A_n$ for some $n \in \omega$, then there is an $m \in \omega$ and a set $\{a_i \mid i \in I\} \subseteq A_m$ such that $s \cup \{\varphi_i(c_{a_i}) \mid i \in I\} \in S$.

(C7) Suppose for some $n \in \omega$, $\{t_i \mid i \in I\}$ is a set of basic terms of $M_{A_n}$ w.r.t. $C_n$.

(a) If $\{a_i, b_i \mid i \in I\} \subseteq A_n$ and $\{c_{a_i} = c_{b_i} \mid i \in I\} \subseteq s$, then $s \cup \{c_{b_i} = c_{a_i} \mid i \in I\} \in S$.

(b) If $\{a_i \mid i \in I\} \subseteq A_n$ and $\{\varphi_i(t_i) \mid i \in I\} \cup \{c_{a_i} = t_i \mid i \in I\} \subseteq s \cap A_n$, then $s \cup \{\varphi_i(c_{a_i}) \mid i \in I\} \in S$.

(c) Then there is an $m \in \omega$ and a set $\{a_i \mid i \in I\} \subseteq A_m$ such that $s \cup \{c_{a_i} = t_i \mid i \in I\} \in S$.

We of course get a model existence theorem for these consistency properties.

2.2 MODEL EXISTENCE THEOREM. *If* S *is an* A *consistency property and* $s \in S$, *then* s *has a model of cardinality less than or equal to the cardinality of* A.

PROOF: As in the proofs of the model existence theorems of the previous chapter, once we have defined a countable increasing sequence of elements of S in which $s = s_0$, we can construct a model of s on the constants of C. So suppose $s_n \in S$ and $T_n = C_n \cup \mathrm{Con}(L_{A_n}) \cup \{\underline{f}(c_{a_1},\ldots,c_{a_k}) \mid k \in \omega,\ \underline{f} \in F_{k,n},\ a_1,\ldots,a_k \in A_n\}$, where $F_{k,n}$ is the set of k place function symbols of $L_{A_n}$. We now let $s_{n+1} = s_n^8$ where we define $s_n^1 - s_n^8$ as we did in the proof of Theorem 1.2(*ii*) of Chapter I except that we consider formulas in $s_n \cap A_n$ instead of in $s_n$. ⊣

Note that for $A \in HC$, the hereditarily countable sets, $L_A \subseteq L_{\omega_1\omega}$ and, since each $A_n$ is finite, any A consistency property is an ordinary consistency property.

3. $\underset{\sim}{\Sigma}_1$ Compactness. In the countable case a $\underset{\sim}{\Sigma}_1$ complete provability predicate is used to prove the Barwise Compactness Theorem. It is not in general possible to find such a provability predicate in the uncountable case. However, if A is a cf $\omega$ $\mathcal{P}$-admissible set we can define a $\underset{\sim}{\Sigma}_1$ in $\mathfrak{D}_{\mathcal{P}}$ on A predicate which will give us the valid sentences of $L_A$.

We define a $\underset{\sim}{\Delta}$ in $\mathfrak{D}_{\mathcal{P}}$ on A function P such that $\varphi \in P(\alpha,a)$ iff $\varphi$ has a proof in $M_A$ of height at most $\alpha$ and such that each step of the proof is in a. I.e., if $\alpha \in A - \mathrm{ord}$, let $P(\alpha,a) = 0$ and if $\alpha \in \mathrm{ord}(A)$, let $P(\alpha,a)$ be the set of all $\varphi \in a \cap M_A$ such that

(1) $\varphi$ is an axiom of $M_A$ (see I.2)

(2) $(\exists\beta\in\alpha)(\exists\psi\in a)(\psi\in P(\beta,a) \wedge (\psi\to\varphi) \in P(\beta,a))$

(3) $(\exists\beta\in\alpha)(\exists\psi,\Theta,g\in a)(\mathrm{fcn}(g) \wedge \mathrm{dm}(g) = \Theta \wedge (\psi\to \bigvee\Theta) \in P(\beta,a) \wedge$

$(\forall\theta,\theta'\in\Theta)(\theta \neq \theta' \to v_{g(\theta)} \notin FV(\psi\wedge\theta')) \wedge \varphi = (\psi\to \bigvee\{\forall v_{g(\theta)}\theta \mid \theta\in\Theta\}))$

or

(4) $(\exists\psi,\Theta\in a)((\forall\theta\in\Theta)\mathrm{Disj}(\theta) \wedge \varphi = (\psi\to \neg\bigwedge\Theta) \wedge$

$(\forall g \in X\{(\theta)_1 \mid \theta\in\Theta\})(\exists\beta\in\alpha)((\psi\to \neg\bigwedge\{g(\theta) \mid \theta\in\Theta\}) \in P(\beta,a))$.

The function $P$ is $\underset{\sim}{\Delta}$ in $\mathcal{D}_{\mathcal{P}}$ on $A$ rather than $\underset{\sim}{\Delta}$ on $A$ because of the bound $X\{(\theta)_1 \mid \theta\in\Theta\}$ in (4). However, if $A \in HC$, we only need (4) for finite $\Theta$ so $P$ will actually be $\underset{\sim}{\Delta}$ on $A$. Note that the conditions (2), (3) and (4) are the rules of inference-modus ponens, rule of disjunctive generalization, and the distributive laws.

We write $\varphi \in P(\alpha,a)$ for a $\underset{\sim}{\Sigma}_1$ (in $\mathcal{D}_{\mathcal{P}}$) definition of $P$ and let $\vdash_A\varphi$ mean $(\exists\alpha,a)(\varphi\in P(\alpha,a))$.

3.1 THEOREM. *If* $A$ *is a* cf $\omega$ $\mathcal{P}$*-admissible set, the set of valid sentences of* $L_A$ *is* $\underset{\sim}{\Sigma}_1$ *in* $\mathcal{D}_{\mathcal{P}}$ *on* $A$. *If* $A \in HC$ *is admissible, the set of valid sentences of* $L_A$ *is* $\underset{\sim}{\Sigma}_1$ *on* $A$.

PROOF. Let $S$ be the set of $s \subseteq \mathrm{Sent}\ M_A$ such that $s \in A$ and not $-\vdash_A \neg\bigwedge s$. Clearly once we know $S$ is an $A$ consistency property we know $\varphi \in L_A$ implies $(\vdash_A\varphi$ iff $\models\varphi)$. Hence $\{\varphi\in L_A \mid \models\varphi\} = \{\varphi\in L_A \mid (\exists\alpha,a\in A)P(\alpha,a)\}$ is $\underset{\sim}{\Sigma}_1$ (in $\mathcal{D}_{\mathcal{P}}$) on $A$.

In order to show $S$ is an $A$ consistency property we first need to know our "new" sets are elements of $A$. As an example we check (C4). Suppose $\Psi = \{\forall v_i\varphi_i(v_i) \mid i\in I\}$, $\Psi \subseteq s\cap A_n$ and $B \subseteq A_n$. If $A \in HC$, $\Psi$ and $B$ are finite subsets of $A$ and hence elements of $A$. If $A$ is $\mathcal{P}$-admissible, $\Psi,B \in \mathcal{P}(A_n)$ and $A_n \in A$ imply $\Psi,B \in A$. We want to show $\Theta \in A$ where $\Theta = \{\varphi_i(c_b) \mid i\in I, b\in B\}$. Clearly $\theta \in \Theta \longleftrightarrow (\exists b\in B)(\exists\psi\in\Psi)(\theta = SF((\psi)_2,(\psi)_1,c_b))$. We know SF is a primitive recursive function, the quantifiers are bounded by elements of $A$ and $A$ is Prim closed, hence $\Theta \in A$. But then since $s \in A$, $s \cup \Theta \in A$.

We check only (C5) and (C6) since (C7c) is similar to (C6) and the others follow directly from the axioms, simple applications of the rules of inference, and the fact that, if $s' \subseteq s$ and $\vdash_A \wedge s \rightarrow \neg \wedge s'$, then $s \notin S$.

(C5) Suppose $\{\vee\Phi_i \mid i \in I\} \subseteq s \cap A_n$ and for all $f \in X\{\Phi_i \mid i \in I\}$, $s \cup \{f(i) \mid i \in I\} \notin S$, i.e.

$$(\forall f \in X\{\Phi_i \mid i \in I\})(\exists \alpha, a \in A)((\wedge s \rightarrow \neg \wedge\{f(i) \mid i \in I\}) \in P(\alpha, a)).$$

By $\underset{\sim}{\Delta}_0$ (in $\mathfrak{F}_{\mathcal{P}}$) collection

$$(\exists b \in A)(\forall f \in X\{\Phi_i \mid i \in I\})(\exists \alpha, a \in b)((\wedge s \rightarrow \neg \wedge\{f(i) \mid i \in I\}) \in P(\alpha, a)).$$

Thus

$$(\exists b \in A)(\forall f \in X\{\Phi_i \mid i \in I\})(\exists \alpha \in \mathrm{ord}(b))((\wedge s \rightarrow \neg \wedge\{f(i) \mid i \in I\}) \in P(\alpha, b))$$

and therefore by rule (4) of the definition of $P$, $(\wedge s \rightarrow \neg \wedge\{\vee\Phi_i \mid i \in I\}) \in P(\mathrm{ord}(b), b)$, i.e. $\vdash_A \wedge s \rightarrow \neg \wedge\{\vee\Phi_i \mid i \in I\}$ and $s \notin S$.

(C6) Suppose $\{\exists v_i \varphi_i(v_i) \mid i \in I\} \subseteq s \cap A_n$, $s \in A_n$ and $\{a_i \mid i \in I\} \subseteq A_{n+1} - A_n$ is a set of $|I|$ distinct elements of $A$. Then for all $i \in I$, $c_{a_i}$ does not appear in $s$. If we let $\varphi_i' = SF(\varphi_i(v_i), v_i, c_{a_i})$, $s \cup \{\varphi_i' \mid i \in I\} \notin S$ implies $\vdash_A \wedge s \rightarrow \vee\{\neg\varphi_i' \mid i \in I\}$, i.e. $(\exists \alpha, a \in A)((\wedge s \rightarrow \vee\{\neg\varphi_i' \mid i \in I\}) \in P(\alpha, a))$. If $b_i = a \times a_i$, no variable $v_{b_i}$ appears in the above proof and if $\varphi_i'' = SF(\varphi_i, v_i, v_{b_i})$, $\varphi_i' = SF(\varphi_i'', v_{b_i}, c_{a_i})$. Since the $v_{b_i}$ are distinct variables and $f = \{(v_{b_i}, c_{a_i}) \mid i \in I\} \in A$ is a function whose range is a set of terms, for any formula $\theta \in M_A$, $SF(\theta, f) \in A$.

Hence $\vdash_A \wedge s \to \bigvee\{\neg\varphi_i' \mid i\in I\}$ iff $\vdash_A SF(\wedge s \to \bigvee\{\neg\varphi_i'' \mid i\in I\}, f)$. The $c_{a_i}$ are also distinct and no $c_{a_i}$ appears in $\wedge s$ or $\bigvee\{\neg\varphi_i'' \mid i\in I\}$ so $\vdash_A \wedge s \to \bigvee\{\neg\varphi_i'' \mid i\in I\}$. We can therefore apply the rule of disjunctive generalization to get $\vdash_A \wedge s \to \bigvee\{\forall v_{b_i} \neg\varphi_i'' \mid i\in I\}$. We can then change bound variables to show $\vdash_A \wedge s \to \bigvee\{\forall v_i \neg\varphi_i \mid i\in I\}$ and therefore $s \notin S$. ⊣

If $A \subseteq HC$ is admissible but not countable, the set of valid sentences of $L_A$ is still $\underset{\sim}{\Sigma}_1$ on $A$ but this cannot be proved using an $A$ consistency property. However this fact can be proved using the obvious ordinary consistency property.

3.2 BARWISE COMPACTNESS THEOREM AND cf $\omega$ COMPACTNESS THEOREM OF BARWISE AND KARP. *Suppose* $A \in HC$ *is admissible or* $A$ *is a* cf $\omega$ $\mathcal{P}$*-admissible set. If* $X$ *is a* $\underset{\sim}{\Sigma}_1$ *on* $A$ *set of sentences of* $L_A$ *such that for every subset* $x$ *of* $X$*,* $x \in A$ *implies* $x$ *has a model, then* $X$ *has a model. (If* $A$ *is* $\mathcal{P}$*-admissible, the same statement holds for* $x$ $\underset{\sim}{\Sigma}_1$ *in* $\mathcal{B}_{\mathcal{P}}$ *on* $A$.*)*

PROOF: Let $S$ be the set of $s \subseteq \text{Sent } M_A$ such that $s$ is $\underset{\sim}{\Sigma}_1$ on $A$, $\text{Con}(s) \cap C \subseteq C_n$ for some $n \in \omega$ and for all $t \subseteq s$, $t \in A$ implies $t$ has a model. Clearly $X$ is an element of $S$ so if $S$ is a consistency property $X$ has a model. Suppose $s \in S$ and $s = \{y\in A \mid \exists w\sigma(w,y)\}$ where $\sigma$ is a $\Delta_0$ formula. Clearly if $a \in A$ then $s \cup a$ is $\underset{\sim}{\Sigma}_1$ on $A$.

(C1) Suppose $\{\varphi, \neg\varphi\} \subseteq s$. Since $\{\varphi,\neg\varphi\} \in A$, it has a model, which is impossible.

(C2) Suppose $\{\neg\varphi \mid \varphi\in\Phi\} \subseteq s \cap A_n$ and $t \subseteq s \cup \{\varphi\neg \mid \varphi\in\Phi\}$ such that $t \in A$. An argument like that at the beginning of the proof of Theorem 3.1 shows that both $\{\neg\varphi \mid \varphi\in\Phi\}$ and $\{\varphi\neg \mid \varphi\in\Phi\}$ are elements of A. But then $(t - \{\varphi\neg \mid \varphi \in \Phi\}) \cup \{\neg\varphi \mid \varphi \in \Phi\}$ is a subset of s which is an element of A. Since it has a model, so must t.

Again we check only (C5) and (C6) since (C7c) is similar to (C6) and all the other cases work precisely as (C2) did.

(C5) Suppose $\{\bigvee\Phi_i \mid i \in I\} \subseteq s \cap A_n$ and for all $f \in X\{\Phi_i \mid i \in I\}$, $s \cup \{f(i) \mid i \in I\} \notin S$. Then for all $f \in X\{\Phi_i \mid i \in I\}$, there is a $t_f \subseteq s \cup \{f(i) \mid i \in I\}$ such that $t_f \in A$ and $t_f$ has no model. Let $t_f' = t_f - \{f(i) \mid i \in I\}$. Then $t_f' \in A$, $t_f' \subseteq s$, and $t_f' \cup \{f(i) \mid i \in I\}$ has no model. Thus we have

$$(\forall f \in X\{\Phi_i \mid i \in I\})(\exists t_f' \in A)(t_f' \subseteq s \wedge \vdash_A \bigwedge t_f' \rightarrow \neg\bigwedge\{f(i) \mid i \in I\}).$$

Since $t_f' \subseteq s$ can be written as $(\forall z \in t_f')(\exists w\sigma(w,z))$, we have, by $\underset{\sim}{\Delta}_0$ collection, an $a \in A$ (which we can assume transitive) such that

$$(\forall f \in X\{\Phi_i \mid i \in I\})(\exists t_f' \in a)((\forall z \in t_f')(\exists w \in a)\sigma(w,z) \wedge \vdash_A \bigwedge t_f' \rightarrow \neg\bigwedge\{f(i) \mid i \in I\}).$$

If $a' = \{z \in a \mid (\exists w \in a)\sigma(w,z)\}$, $a'$ is a $\underset{\sim}{\Delta}_0$ subset of a and hence an element of A. But for all $f \in X\{\Phi_i \mid i \in I\}$, $t_f' \subseteq a'$, and therefore for all $f \in X\{\Phi_i \mid i \in I\}$, $\vdash_A \bigwedge a' \rightarrow \neg\bigwedge\{f(i) \mid i \in I\}$, i.e., $a' \cup \{f(i) \mid i \in I\}$ has no model. But then $a' \cup \{\bigvee\Phi_i \mid i \in I\}$ has no model, is a subset of s, and is an element of A, a contradiction.

(C6) Suppose $\{\exists v_i\varphi_i(v_i) \mid i \in I\} \subseteq s \cap A_n$. Choose new constants $\{c_{a_i} \mid i \in I\}$ as we did in (C6) of the proof of Theorem 3.1 and suppose $s' = s \cup \{\varphi_i(c_{a_i}) \mid i \in I\}$, $t \subseteq s'$, and $t \in A$. Let $t' = t - \{\varphi_i(c_{a_i}) \mid i \in I\}$. Then $t' \in A$ and $t' \subseteq s$. If $t'' = t' \cup \{\exists v_i\varphi_i(v_i) \mid i \in I\}$, $t'' \in A$ and $t'' \subseteq s$, hence $t''$ has a model. But no $c_{a_i}$ appear in $t''$ so we can clearly find a model of t. ⊣

Although we include no applications of Theorem 3.2 here, we note that all next $\mathcal{P}$-admissible sets are cf $\omega$ $\mathcal{P}$-admissible sets [10] so that many of the applications of the Barwise compactness theorem which use next admissible sets can be generalized to be applications of the cf $\omega$ compactness theorem of Barwise and Karp.

## REFERENCES


1. J. Barwise, *Infinitary logic and admissible sets*, Journal of Symbolic Logic 34(1969), 226-252.

2. J. Barwise, *Applications of strict* $\Pi_1^1$ *predicates to infinitary logic*, Journal of Symbolic Logic 34(1969), 409-423.

3. J. Barwise and K. Kunen, *Hanf numbers for fragments of* $L_{\infty\omega}$, Israel Journal of Math. 10(1971), 306-320.

4. C.C. Chang, *Two interpolation theorems*, Istituto Nazionale di Alta Mathematica, Symposia Mathematica V(1970), 5-19.

5. C.C. Chang, *Some remarks on the model theory of infinitary languages*, in: The Syntax and Semantics of Infinitary Languages, Springer-Verlag, Berlin, 1968, 36-63.

6. E. Cunningham, *Chain Models for Infinite-Quantifier Languages*, Ph.D. Thesis, University of Maryland, 1974.

7. E. Cunningham, *Chain models: applications of consistency properties and back-and-forth techniques in infinite-quantifier languages*, this volume.

8. H. Friedman, *The Beth and Craig theorems in infinitary languages*, (mimeographed).

9. J. Green, $\Sigma_1$ *compactness for next admissible sets*, Journal of Symbolic Logic 39(1974), 105-116.

10. J. Green, *Next* $\mathcal{P}$ *admissible sets are of cofinality* $\omega$, to appear.

11. J. Gregory, *Beth definability in infinitary languages*, Journal of Symbolic Logic 39(1974), 22-26.

12. M. Helling, *Hanf numbers for some generalizations of first-order languages*, Notices of the American Mathematical Society 11(1964), 679.

13. R. Jensen and C. Karp, *Primitive recursive set functions*, in: Axiomatic Set Theory, part 1, American Mathematical Society, 1971, 143-176.

14. C. Karp, Languages with Expressions of Infinite Length, North-Holland, Amsterdam, 1964.

15. C. Karp, *Nonaxiomatizability results for infinitary systems*, Journal of Symbolic Logic 32(1967), 367-384.

16. C. Karp, *An algebraic proof of the Barwise compactness theorem*, in: The Syntax and Semantics of Infinitary Languages, Springer-Verlag, Berlin, 1968, 80-95.

17. C. Karp, *Infinite-qunatifier languages and* $\omega$*-chains of models*, Proceedings of the Tarski Symposium, American Mathematical Society, 1974, 225-232.

18. H.J. Keisler, Model Theory for Infinitary Logic, North-Holland, Amsterdam, 1971.

19. A. Lévy, *A hierarchy of formulas in set theory*, Memoirs of the American Mathematical Society, No. 57, 1965.

20. E.G.K. López-Escobar, *An interpolation theorem for denumerably long sentences*, Fundamenta Mathematica LVII(1965), 253-272.

21. E.G.K. López-Escobar, *On defining well-orderings*, Fundamenta Mathematica LIX(1966), 13-21.

22. M. Makkai, *An application of a method of Smullyan to logics on admissible sets*, Bulletin de l'Académie Polonaise des Sci, Ser. Math. 17(1969), 341-346.

23. J. Malitz, *Problems in the Model Theory of Infinitary Languages*, Ph.D. Thesis, Berkeley, 1966.

24. R. Platek, *Foundations of Recursion Theory*, Ph.D. Thesis, Stanford, 1966.

25. R. Smullyan, *A unifying principle in quantification theory*, Proceedings of the National Acad. Sci. 49(1963), 828-832.

PART C

# CHAIN MODELS:

# APPLICATIONS OF CONSISTENCY PROPERTIES AND BACK-AND-FORTH TECHNIQUES IN INFINITE-QUANTIFIER LANGUAGES

BY

ELLEN CUNNINGHAM

# CONTENTS
# PART C

1. Historical Introduction. In her doctoral dissertation, while investigating the independence of various axiom systems for infinitary languages, Karp introduced the idea of a *general model*, which consists of a universe $A$, interpretations of relation and constant symbols, and a specified set $S$ of functions from sets of variables into $A$, satisfying certain closure conditions (see [6]). A standard model can be viewed as a general model in which $S$ happens to consist of all functions from sets of variables into $A$.

Equivalently, a general model may be defined as a directed system of models -- this is the formulation which Karp introduced in [5] and which she preferred as the more natural one. The simplest non-standard example is a sequence $\underset{\sim}{\mathfrak{A}} = \langle \mathfrak{A}_n \mid n \in \omega \rangle$ of standard models such that $\mathfrak{A}_n \subseteq \mathfrak{A}_{n+1}$ for each $n$ (though $\mathfrak{A}_{n+1}$ is allowed to interpret more constants than $\mathfrak{A}_n$), in which satisfaction of infinite-quantifier formulas is defined only for assignments to variables which are bounded by some $\mathfrak{A}_n$. The definition of satisfaction is the standard one for atomic formulas and over propositional connectives. Over the existential quantifier the definition is as follows: If $\vec{a}$ is a function from the free variables of $(\exists V)\varphi$ to some $A_n$, then $\underset{\sim}{\mathfrak{A}} \models (\exists V)\varphi[\vec{a}]$ iff there is some $m \in \omega$ and some $\vec{b}: V \longrightarrow A_m$ such that $\underset{\sim}{\mathfrak{A}} \models \varphi[\vec{a},\vec{b}]$. Such a $\vec{b}$ is referred to as a *bounded assignment*. Since $(\forall V)$ can be defined as $\neg(\exists V)\neg$, satisfaction over the universal quantifier $(\forall V)$ involves all bounded assignments to $V$.

Karp called such sequences $\underset{\sim}{\mathfrak{A}}$, with satisfaction of formulas defined as above, $\omega$-chains of models (see [4]). Following Makkai [9], we refer to them simply as *chain models*. $\underset{\sim}{\mathfrak{A}}$ and $\underset{\sim}{\mathfrak{B}}$ will denote $\langle \mathfrak{A}_n \mid n \in \omega \rangle$ and $\langle \mathfrak{B}_n \mid n \in \omega \rangle$ respectively. Note that any standard model $\mathfrak{A}$ can be considered as the chain model $\langle \mathfrak{A}, \mathfrak{A}, \mathfrak{A}, \ldots \rangle$. Thus if a sentence $\varphi$ is valid in chain models (written $\models^{\omega} \varphi$) it is valid in standard models (written $\models \varphi$ as usual).

For finite-quantifier $\varphi$ the converse is also true: $\models \varphi$ implies $\models^{\omega} \varphi$, since every finite assignment is bounded.

In his lecture notes on the model theory of infinitary languages (now in book form [7]) Keisler showed the usefulness of consistency properties for $L_{\omega_1\omega}$. Karp's student Judy Green extended these investigations by defining consistency properties and proving the corresponding model existence theorems for certain uncountable finite-quantifier languages, including $L_{\kappa\omega}$ for $\kappa$ strong limit of cofinality $\omega$ (see the article by Green elsewhere in this volume, and also [3]).

In this article $\kappa$ is always used for such a cardinal and we assume $\kappa > \omega$ unless stated otherwise. We write $\kappa = \bigcup_{n\in\omega} \kappa_n$ where $\kappa_n < \kappa_{n+1} < \kappa$ for all $n$.

While Green's work was in progress, Karp observed that if her definition of $\kappa$ consistency properties for $L_{\kappa\omega}$ were extended in a natural way to $L_{\kappa\kappa}$, the result was the existence of a chain model, though not necessarily of a standard model. A chain model is said to be of *strict power* $\kappa$ if $|\bigcup_{n\in\omega} A_n| = \kappa$ and $|A_n| < \kappa$ for each $n$. $\underset{\sim}{\mathfrak{A}}$ is said to be of *strict power at most* $\kappa$ if $|A_n| < \kappa$ for each $n$. So every chain model with $|\bigcup_{n\in\omega} A_n| < \kappa$ has strict power at most $\kappa$. In [4] Karp presented some of the early evidence that chain models of strict power $\kappa$ are in many ways as well-behaved toward the languages $L_{\kappa\kappa}$ as are countable models toward $L_{\omega_1\omega}$. (Note that if $\kappa > \omega$ then $L_{\kappa\kappa}$ is equivalent in expressive power to $L_{\kappa^+\kappa}$.)

The investigation of chain models for $L_{\kappa\kappa}$ was continued by the present writer in her doctoral dissertation [2]. It is the purpose of this article to explain in brief how Karp's techniques were modified and applied to answer some of her conjectures and in some cases to improve on her results. Full details are given in [2]. Some familiarity with Green's work, say as presented in this volume, is presumed.

2. Chain Consistency Properties. The similarity between the following definition and those of Keisler, Green, and Karp is very noticeable.

Definition 2.1. Let $\sigma$ be a similarity type, which may include relation and constant symbols, but no function symbols. Let $C = \bigcup\{C_n \mid n\in\omega\}$ be a set of constants disjoint from $\sigma$, and such that $|C_n| = \kappa_n$ and $C_n \subseteq C_{n+1}$ for all $n$. Let $\underset{\sim}{C}$ denote the sequence $\langle C_n \mid n\in\omega\rangle$.

A collection $S$ of sets of sentences of $L_{\kappa\kappa}(\sigma \cup C)$ is a *chain consistency property for* $L_{\kappa\kappa}(\sigma)$ *with respect to* $\underset{\sim}{C}$ if, for each $s \in S$, $|s| \leq \kappa$ and all of the following hold:

(C1) For atomic $\alpha$, if $\alpha \in s$ then $\neg\alpha \notin s$.

(C2) If $\{\neg\varphi \mid \varphi\in\Phi\} \subseteq s$ and $|\Phi| < \kappa$ then $s \cup \{\varphi\neg \mid \varphi\in\Phi\} \in S$.

(C3) If $\{\wedge\Phi_i \mid i\in I\} \subseteq s$, $|I| < \kappa$, and for some $m$, $|\Phi_i| \leq \kappa_m$ for all $i \in I$, then $s \cup \bigcup\{\Phi_i \mid i\in I\} \in S$.

(C4) If $\{(\forall V_i)\varphi_i \mid i\in I\} \subseteq s$, $|I| < \kappa$, and for some $m$, $|V_i| \leq \kappa_m$ for all $i \in I$, then for every $n$,
$s \cup \{\varphi_i(V_i/f_i) \mid i\in I,\ f_i\colon V_i \longrightarrow C_n\} \in S$.

(C5) If $\{\vee\Phi_i \mid i\in I\} \subseteq s$, $|I| < \kappa$, and for some $m$, $|\Phi_i| \leq \kappa_m$ for all $i \in I$, then there is $f \in \times\{\Phi_i \mid i\in I\}$ such that $s \cup \{f(i) \mid i\in I\} \in S$.

(C6) If $\{(\exists V_i)\varphi_i \mid i\in I\} \subseteq s$, $|I| < \kappa$, and for some $m$, $|V_i| \leq \kappa_m$

for all $i \in I$, then there is a function $\underline{n}: I \longrightarrow \omega$, and functions $f_i: V_i \longrightarrow C_{\underline{n}'i}$ for each $i \in I$, such that $s \cup \{\varphi_i(V_i/f_i) \mid i \in I\} \in S$.

(C7) (a) If $\{c_i = c_i' \mid i \in I\} \subseteq s$ and $|I| < \kappa$, where the $c_i$ and $c_i'$ are elements of $C$, then $s \cup \{c_i' = c_i \mid i \in I\} \in S$.

(b) If $\bigcup_{i \in I} \{\alpha(e_i), e_i = c_i\} \subseteq s$, $\alpha$ atomic or negated atomic, $\{c_i \mid i \in I\} \subseteq C_n$ for some $n$, $e_i \in \sigma \cup C$, $|I| < \kappa$, then $s \cup \{\alpha(c_i) \mid i \in I\} \in S$.

(c) If $\{e_i \mid i \in I\}$ is a set of constants in $\sigma \cup C$, $|I| < \kappa$, then there are $c_i$ in $C$ such that $s \cup \{e_i = c_i \mid i \in I\} \in S$.

We have the following

2.2 THEOREM. *A sentence* $\varphi$ *of* $L_{\kappa\kappa}(\sigma)$ *is true in a chain model iff* $\{\varphi\}$ *belongs to a chain consistency property.*

PROOF: The proof from right to left is a Henkin-style construction, modeled on those of Green [3]. The idea is to construct a chain $\langle s_n \mid n \in \omega \rangle$ of members of $S$ such that $\{\varphi\} = s_0$, and then to define a chain model $\underset{\sim}{\mathfrak{A}}$ as follows: For $c \in C$ we define $c \sim c'$ iff the sentence $c = c'$ is in $\bigcup_n s_n$. The $n^{\underline{th}}$ universe of $\underset{\sim}{\mathfrak{A}}$ is the set of equivalence classes $c^\sim$ of the $c \in C_n$. If $e$ is an individual constant in $\sigma$ we let $e^{\mathfrak{A}}$ be $c^\sim$ provided the sentence $c = e$ is in $\bigcup_n s_n$, and if $R$ is a k-ary predicate symbol in $\sigma$ we say $\langle c_1^\sim, \ldots, c_k^\sim \rangle \in R^{\mathfrak{A}}$ provided $R(c_1, \ldots, c_n)$ is in $\bigcup_n s_n$. Of course, the sets $s_n \in S$ are carefully chosen to insure that $\sim$ is an equivalence relation and that the interpretations of constants and predicate symbols are well-defined, and also to show that the resultant chain model $\underset{\sim}{\mathfrak{A}}$ satisfies every sentence in $\bigcup_n s_n$, in particular the given sentence $\varphi$.

The converse, as will be explained below, is more delicate. It is an immediate consequence of the downward Löwenheim-Skolem theorem for chain models (Corollary 2.5, below) and the easy Lemma 2.3. ⊣

2.3 LEMMA. *If* $\varphi$ *is a sentence of* $L_{\kappa\kappa}(\sigma)$ *which is true in a chain model* $\underset{\sim}{\mathfrak{A}}$ *of strict power at most* $\kappa$, *then* $\{\varphi\}$ *belongs to a chain consistency property.*

PROOF: For each $n$, let $p_n$ be the least integer $p > p_{n-1}$ such that $|A_n| \leq \kappa_p$. Such a $p$ exists since $|A_n| < \kappa$ for all $n$. We expand $\underset{\sim}{\mathfrak{A}}$ to a chain model $\underset{\sim}{\mathfrak{A}}'$ for $\sigma \cup C$ so that $A_n$ is exactly the set of interpretations of the constants of $C_{p_n}$. This can be done since $|C_{p_n}| = \kappa_{p_n}$. Then let $S$ be the collection of sets $s$ of sentences of $L_{\kappa\kappa}(\sigma \cup C)$ such that $\mathfrak{A}' \models \wedge s$. Clearly $\{\varphi\} \in S$, and the conditions of Definition 2.1 are easily verified. ⊣

Karp's original definition of a $\kappa$ consistency property for $L_{\kappa\kappa}(\sigma)$ w.r.t. $\underset{\sim}{C}$ is like Definition 2.1 except in clauses (C6), where the function $\underline{n}$ is required to be constant, and (C7)(c) where the $c_i$ for $i \in I$ are required to all belong to some $C_m$. Her definition yields a chain model existence theorem, like the right-to-left direction of Theorem 2.2, by essentially the same proof. The "easy" left-to-right part fails in this case, however; a sentence may be true in a chain model, even one of strict power $\kappa$, and yet not belong to a $\kappa$ consistency property in Karp's sense. The classic example is a sentence of the form

$$\bigwedge_{n\in\omega} (\exists v_n)\varphi_n(v_n) \wedge \neg(\exists\{v_n \mid n\in\omega\}) \bigwedge_{n\in\omega} \varphi_n(v_n),$$

i.e., a negation of the axiom of independent choices. This is not surprising since the $\kappa$ consistency properties seem to correspond to the axiom system which Karp called $\kappa$-logic, and chain models were first used by her precisely to show that the axiom of independent choices was independent of the other axioms of that system (namely the axioms of ordinary predicate logic and the $\kappa_n$ distributive laws.)

The difficulty with such a sentence arises precisely in clause (C6) if the function $\underline{n}$ is required to be bounded by some $C_m$. For in the chain models constructed from consistency properties, the denotations of the $c \in C_m$ comprise the $m^{\underline{th}}$ universe, and thus no bounded set of $c$'s could witness to the $v_n$'s in this case. Karp was undoubtedly aware that a definition such as 2.1 would make exact the correspondence between truth in a chain model and membership in a consistency property. Her reason for nonetheless insisting on a bounded function $\underline{n}$ was to strengthen the induction hypothesis needed to verify clause (C6) in applications, by insuring that there was always a sufficient supply of "new" witnessing constants available. This was the device used successfully by Green [3] in proving the completeness theorem and interpolation theorems for $L_{\kappa\omega}$ and so it was very natural to adapt it to $L_{\kappa\kappa}$. In each of Green's proofs a set $S$ is defined by some sort of consistency condition, and when verifying (C6) one has to show, for some set $\{(\exists V_i)\varphi_i \mid i\in I\} \subseteq s \in S$, that there are $n \in \omega$ and mappings $f_i\colon V_i \longrightarrow C_n$ for $i \in I$, such that $s \cup \{\varphi_i(V_i/f_i) \mid i\in I\} \in S$. One assumes that the $c$'s occurring in $s$ all belong to some $C_m$, and can thus choose $n$ large enough so that the $f_i(v)$ can be chosen new and distinct. But if $s \cup \{\varphi_i(V_i/f_i) \mid i\in I\}$ does not satisfy the consistency condition defining $S$, then by an application of the rule of disjunctive generalization, neither does $s$, contradicting the assumption $s \in S$.

In applying Green's method to $L_{\kappa\kappa}$, however, Karp encountered a further difficulty: the rule of disjunctive generalization, since it is dual to the axiom of independent choices, is valid in standard models but not in chain models. Thus, for example, Karp's version of the Craig interpolation theorem, as given in [4], has a weaker conclusion than one would like, namely, for an implication $\varphi_1 \longrightarrow \varphi_2$ valid in chain models one finds an interpolant $\theta$ in the common languages such that $\varphi_1 \longrightarrow \theta$ and $\theta \longrightarrow \varphi_2$ are valid in *standard* models.

The foregoing seemed to suggest that, in working with chain models, better results could be obtained if one had suitably restricted versions of the axiom of independent choice and the rule of disjunctive generalization. Following a suggestion of E.G.K. López-Escobar, the author augmented the languages $L_{\kappa\kappa}(\sigma)$ by a set $B = \{B_n \mid n \in \omega\}$ of special unary "bounding predicates," introducing the concept of a *B-model* (i.e., a chain model in which $B_n$ is interpreted as the $n^{\underline{th}}$ universe), and formulated a system of B-axioms, which are valid in all B-models.

All but a few of the B-axiom schemata and rules are borrowed from $\kappa$-logic. An example of an exception is the following rule, a modified version of the rule of disjunctive generalization:

From $\psi \longrightarrow \bigvee_{i \in I} \bigwedge_{n \in \omega} (B_n(f''V_i) \longrightarrow \varphi_i(V_i/f_i))$, infer $\psi \longrightarrow \bigvee_{i \in I} (\forall V_i)\varphi_i(V_i)$, where $f$ is a one-one mapping of $\bigcup_{i \in I} V_i$ into a set of constants not occurring in the conclusion, and for every $i$, no $v \in V_i$ occurs in any other quantifier in the conclusion and no $v \in V_i$ occurs free in the conclusion.

Not only are the B-axioms and rules (and thus all theorems of the system) valid in B-models, but a sentence of $L_{\kappa\kappa}(\sigma)$ (i.e., not containing $B_n$'s) is valid in B-models iff it is valid in all chain models. More importantly, we have the following.

2.4 THEOREM (Completeness theorem). *Let* $\varphi$ *be a sentence of* $L_{\kappa\kappa}(\sigma)$ *which is consistent with respect to the* B-*axioms and rules. Then* $\varphi$ *has a chain model of strict power at most* $\kappa$.

PROOF: To prove this we define a chain consistency property $S$ such that, for every $s \in S$, $s$ is consistent (in the sense of B-provability) with the sentence

$$\beta(C^s): \bigwedge_{n\in\omega} \bigwedge\{B_n(c) \mid c\in C_n \text{ and } c \text{ occurs in } s\}.$$

To circumvent the difficulty mentioned above, of having "enough" witnessing constants to verify (C6) and (C7)(c), we employ the following device: each $C_n$ is partitioned into $n$ disjoint sets $C_{n0}, C_{n1}, \ldots, C_{nn}$, each of power $\kappa_n$, in such a way that $C_{nk} \subseteq C_{n+1,k}$ for all $n$, all $k \le n$. We then define $G_m = \bigcup\{C_{nk} \mid n \in \omega, k \le m\}$, so that $C_m \subseteq G_m$, and require that the set of $c$'s occurring in each $s$ be of power $< \kappa$ and bounded by some $G_m$.

We shall outline the verification of clause (C6), the most difficult part of the proof that $S$ is a chain consistency property.

Let $\{(\exists V_i)\varphi_i \mid i\in I\} \subseteq s$, where $|I| \le \kappa_m$, $|V_i| \le \kappa_m$ for all $i$. The constants of $s$ are all in some $G_{k-1}$; we may assume $k > m$. We claim there is some $\underline{n}: I \longrightarrow (\omega \sim k)$ and some set of functions $f_i^{\underline{n}}: V_i \longrightarrow C_{\underline{n}`i}$, for $i \in I$, such that

(1) $$s \cup \{\varphi_i(V_i/f_i^{\underline{n}}) \mid i\in I\} \in S.$$

We shall suppose this fails for all such $\underline{n}$, and obtain a contradiction. Let $\underline{n}: I \longrightarrow (\omega\sim k)$ be fixed. Now for $n \ge k$, $|C_{nk}| \ge \kappa_k > \kappa_m = |\bigcup_{i\in I} V_i|$. Thus we can find one-one functions $f_i^{\underline{n}}: V_i \longrightarrow C_{\underline{n}`i,k}$ such that $Rg(f_i^{\underline{n}}) \cap Rg(f_j^{\underline{n}}) = \phi$ if $i \ne j$. Note that $C_{nk} \subseteq G_k \sim G_{k-1}$ for every $n \ge k$ and thus no $c$ in the range of any $f^{\underline{n}}_i$ occurs in s. $\cup_{i\in I} f^{\underline{n}}_i$ is a function; let it be denoted by $f^{\underline{n}}$. Now if (1) fails for this particular $\underline{n}$, it is because

$$(2) \quad \vdash^{B} \wedge s \wedge \beta(C^{s}) \longrightarrow \left[\bigwedge_{i\in I} B_{\underline{n}^{\prime}i}(f^{\underline{n}\,\prime\prime}V_i) \longrightarrow \neg \bigwedge_{i\in I} \varphi_i(V_i/f^{\underline{n}})\right],$$

by the definition of S.

We now let g be an appropriate mapping to a set of new constants, and replace $f^{\underline{n}}$ by g throughout the proof. This and a tautology applied to (2) yield

$$(3) \quad \vdash^{B} \wedge s \wedge \beta(C^{s}) \longrightarrow \bigvee_{i\in I} (B_{\underline{n}^{\prime}i}(g^{\prime\prime}V_i) \longrightarrow \neg\varphi_i(g^{\prime\prime}V_i)).$$

Now since (3) holds for every $\underline{n}\colon I \to (\omega\sim k)$, we have, after an application of the distributive law:

$$\vdash^{B} \wedge s \wedge \beta(C^{s}) \longrightarrow \bigvee_{i\in I} \bigwedge_{n\geq k} (B_n(g^{\prime\prime}V_i) \longrightarrow \neg\varphi_i(g^{\prime\prime}V_i)).$$

It can be shown that the latter is a sufficient hypothesis for the application of the rule of disjunctive generalization, and so we have $\vdash^{B} \wedge s \wedge \beta(C^{s}) \longrightarrow \vee_{i\in I} (\forall V_i)\neg\varphi_i(V_i)$. However, $\vdash^{B} \wedge s \wedge \beta(C^{s}) \longrightarrow \wedge_{i\in I} (\exists V_i)\varphi_i(V_i)$, so we have the contradiction we sought. ⊣

As an immediate consequence we have a downward Löwenheim-Skolem theorem.

2.5 COROLLARY. *If* $\varphi$ *is a sentence of* $L_{\kappa\kappa}(\sigma)$ *which is true in some chain model, then it is true in some chain model of strict power at most* $\kappa$.

The B-predicates enable us to state various other consistency conditions with slightly more subtlety. We are thus able to verify that certain sets of sentences are chain consistency properties, although they are not $\kappa$ consistency properties in Karp's sense.

Thus, for example, we have the desired strengthening of the Craig interpolation theorem:

2.6 THEOREM. *Let* $\varphi$ *be a sentence of* $L_{\kappa\kappa}(\sigma_1)$*,* $\psi$ *a sentence of* $L_{\kappa\kappa}(\sigma_2)$*, and suppose* $\models^{\omega} \varphi \rightarrow \psi$*. Then there is a sentence* $\theta$ *of* $L_{\kappa\kappa}(\sigma_1 \cap \sigma_2)$ *such that* $\models^{\omega} (\varphi \rightarrow \theta) \wedge (\theta \rightarrow \psi)$.

PROOF: Define S to be the collection of all sets of sentences s such that $|s| \leq \kappa$ and s can be written as $s_1 \cup s_2$, where

(*i*) For $j = 1,2$, $s_j \subseteq L_{\kappa\kappa}(\sigma_j \cup G_k)$ for some $k \in \omega$,

(*ii*) s contains fewer than $\kappa$ $c \in C$,

(*iii*) There is no sentence $\theta$ of $L_{\kappa\kappa}((\sigma_1 \cap \sigma_2) \cup C)$ such that $\wedge s_1 \wedge \beta \rightarrow \theta$ is valid in all B-models, $\wedge s_2 \wedge \beta \rightarrow \neg\theta$ is valid in all B-models, and every c in $\theta$ occurs in both $s_1$ and $s_2$.

(Here $\beta$ is the sentence $\wedge\{B_m(c) \mid c \in C_m, m \in \omega\}$.)
We show that S is a chain consistency property w.r.t. $\underset{\sim}{C}$. Then $\{\varphi, \neg\psi\}$ cannot belong to S since there is no chain model of $\varphi \wedge \neg\psi$, whereupon the existence of the interpolant $\theta$ follows from (*iii*). The details are omitted. ⊣

As a corollary to the theorem we have the result of Chang [1] for standard models:

2.7 COROLLARY. *Let* $\varphi$ *and* $\psi$ *be sentences of* $L_{\kappa\omega}(\sigma_1)$*,* $L_{\kappa\omega}(\sigma_2)$ *respectively, and suppose* $\models \varphi \rightarrow \psi$*. Then there is a sentence* $\theta$ *of* $L_{\kappa\kappa}(\sigma_1 \cap \sigma_2)$ *such that* $\models (\varphi \rightarrow \theta) \wedge (\theta \rightarrow \psi)$.

PROOF: Since $\varphi$ and $\psi$ are finite-quantifier sentences, we actually have $\models^{\omega} \varphi \rightarrow \psi$. So, by the theorem, there is a $\theta$ in

$L_{\kappa\kappa}(\sigma_1 \cap \sigma_2)$ such that the sentence $(\varphi \rightarrow \theta) \wedge (\theta \rightarrow \psi)$ is valid in chain models; in particular it is valid in standard models. $\dashv$

Chain model versions of the Lyndon and Malitz interpolation theorems are proved like Theorem 2.6, and the corresponding preservation theorems (for chain homomorphisms and chain submodels) follow.

3. Chain Isomorphism and Equivalence. Besides consistency properties, another useful device in the model theory of $L_{\omega_1\omega}$ is the back-and-forth argument. Karp observed that just as the best results with $L_{\omega_1\omega}$ are obtained for countable models, so the best results with $L_{\kappa\kappa}$ are obtained for chain models of strict power $\kappa$. For chain models $\underset{\sim}{\mathfrak{A}}$ and $\underset{\sim}{\mathfrak{B}}$, and for any cardinal $\lambda$, $\lambda$-back-and-forth relations between bounded sequences of elements of length less than $\lambda$ are defined just as they are in the standard model case (see, for example, [1] or Kueker's paper in this volume), and the standard characterization of $L_{\infty\lambda}$-equivalence becomes a corollary of this more general result:

3.1 THEOREM. *For chain models* $\underset{\sim}{\mathfrak{A}}$ *and* $\underset{\sim}{\mathfrak{B}}$, $\underset{\sim}{\mathfrak{A}} \equiv_{\infty\lambda} \underset{\sim}{\mathfrak{B}}$ *iff there is a* $\lambda$*-back-and-forth relation between sequences from* $\underset{\sim}{\mathfrak{A}}$ *and* $\underset{\sim}{\mathfrak{B}}$ *(written* $\underset{\sim}{\mathfrak{A}} \cong_\lambda \underset{\sim}{\mathfrak{B}}$*).*

For any chain model $\underset{\sim}{\mathfrak{A}}$, one can write down the $\lambda$-back-and-forth conditions in an appropriate infinitary language, i.e., there is a sentence $\tau_{\underset{\sim}{\mathfrak{A}}}$ such that for any $\underset{\sim}{\mathfrak{B}}$, $\underset{\sim}{\mathfrak{B}} \models \tau_{\underset{\sim}{\mathfrak{A}}}$ iff $\underset{\sim}{\mathfrak{B}} \cong_\lambda \underset{\sim}{\mathfrak{A}}$. If $\mathfrak{A}$ is a standard model, the sentence $\tau_{<\mathfrak{A},\mathfrak{A},\ldots>}$ is just the usual Scott sentence $\tau_{\mathfrak{A}}$ of $\mathfrak{A}$ (see [1]). However, by Theorem 3.1, $\tau_{\mathfrak{A}}$ characterizes $\mathfrak{A}$ up to $L_{\infty\lambda}$-equivalence not only among standard models but among all chain models.

If $\mathfrak{A}$ is a standard model of power $\kappa$ (as before, $\kappa$ is a strong limit cardinal of cofinality $\omega$) then the sentence $\tau_{\mathfrak{A}}$ defining $\mathfrak{A}$ up to $L_{\infty\kappa}$-equivalence will in general be in $L_{\kappa^*\kappa}$ (where $\kappa^{\mathfrak{E}} = \bigcup_{\lambda<\kappa} \kappa^{\lambda}$ and $\kappa^* = (\kappa^{\mathfrak{E}})^+$). However, if $\underset{\sim}{\mathfrak{A}}$ is a chain model of strict power $\kappa$, since the number of bounded sequences from $\underset{\sim}{\mathfrak{A}}$ is $\kappa$, $\tau_{\underset{\sim}{\mathfrak{A}}}$ is in $L_{\kappa\kappa}$. It is then easy to show

3.2 THEOREM (Scott Isomorphism Theorem). *If* $\underset{\sim}{\mathfrak{A}}$ *and* $\underset{\sim}{\mathfrak{B}}$ *are of strict power* $\kappa$ *and* $\underset{\sim}{\mathfrak{A}} \equiv_{\kappa\kappa} \underset{\sim}{\mathfrak{B}}$, *then* $\underset{\sim}{\mathfrak{A}} \cong \underset{\sim}{\mathfrak{B}}$.

(Between chain models $\underset{\sim}{\mathfrak{A}}$ and $\underset{\sim}{\mathfrak{B}}$ we consider only *bounded isomorphisms*, i.e., those isomorphsims $f\colon \bigcup_{n\in\omega} \mathfrak{A}_n \longrightarrow \bigcup_{n\in\omega} \mathfrak{B}_n$ such that $f(A_n) \subseteq B_m$ for some $m$ and $f^{-1}(B_n) \subseteq A_m$ for some $m$. This restriction is easy to justify when we regard chain models as general models.)

For a standard model $\mathfrak{A}$, if $\mathfrak{A} = \bigcup_{n\in\omega} \mathfrak{A}_n$ for some sequence $\langle \mathfrak{A}_n \mid n\in\omega \rangle$ then $\underset{\sim}{\mathfrak{A}} = \langle \mathfrak{A}_n \mid n\in\omega \rangle$ is called a *decomposition* of $\mathfrak{A}$. Karp was very interested in the problem of characterizing those standard models of power $\kappa$ which have $L_{\kappa\kappa}$-equivalent decompositions of strict power $\kappa$. (Note in the countable case, a standard model of power $\omega$ is $L_{\omega_1\omega}$-equivalent to *any* strict-power-$\omega$ decomposition.) The following theorem provides one answer to the question.

3.3 THEOREM. *Let* $\mathfrak{A}$ *be a standard model of power* $\kappa$. *Then the following are equivalent:*

(1) $\mathfrak{A}$ *has a decomposition* $\underset{\sim}{\mathfrak{A}}$ *of strict power* $\kappa$ *such that* $\mathfrak{A} \equiv_{\infty\kappa} \underset{\sim}{\mathfrak{A}}$.

(2) *The* $L_{\infty\kappa}$-*equivalence class of* $\mathfrak{A}$ *is defined by a sentence of* $L_{\kappa\kappa}$.

(3) *If we define* $\vec{a} \approx \vec{b}$ *iff* $(\mathfrak{A},\vec{a}) \equiv_{\kappa\kappa} (\mathfrak{A},\vec{b})$ *for every* $\vec{a}, \vec{b}$ *in* $A^{\gamma}$, *every* $\gamma < \kappa$, *then the total number of equivalence classes under* $\approx$ *is at most* $\kappa$.

(1) $\Longrightarrow$ (2) is easily proved from the preceding results: we take $\tau_{\mathfrak{A}}$ to be the desired sentence. The proof of (2) $\Longrightarrow$ (1) uses the downward Löwenheim-Skolem theorem for chain models. (1) $\Longrightarrow$ (3) and (3) $\Longrightarrow$ (1) are proved in a straightforward manner.

The preceding theorem may be interpreted as saying that the models which have equivalent strict-power-$\kappa$ decompositions are in some sense simpler than those which do not have such decompositions. Another criterion for simplicity is the number of automorphisms a model has -- the more automorphisms, the simpler the structure. Here, too, the models with equivalent strict-power-$\kappa$ decompositions qualify as simple, since they have the largest possible number of automorphisms. This follows from the chain model analogue of a theorem of Kueker's, connecting the number of automorphisms of a countable model with a definability condition [8]. In the chain model case, the "definability" is very loose, but is sufficient to give the desired result.

3.4 THEOREM. *If* $\mathfrak{A}$ *is a chain model of strict power* $\kappa$, *the following are equivalent:*

(1) $\mathfrak{A}$ *has* $\leq \kappa$ *automorphisms.*

(2) $\mathfrak{A}$ *has* $< 2^{\kappa}$ *automorphisms.*

(3) *For some bounded sequence* $\vec{a}$ *of elements of* $\mathfrak{A}$, *and some* $m \in \omega$, *if* $\{\vec{x}_{\xi} \mid \xi < \kappa_m\}$ *is a set of bounded sequences such that* $(\mathfrak{A},\vec{a},\vec{x}_{\xi}) \cong (\mathfrak{A},\vec{a},\vec{x}_{\zeta})$ *for* $\xi < \zeta < \kappa_m$, *then* $\vec{x}_{\xi} = \vec{x}_{\zeta}$ *for some* $\xi \neq \zeta$.

(4) *There are formulas* $\varphi_{\nu,n}(\vec{v},u_0,\ldots,u_\xi,\ldots)_{\xi<\kappa_n}$ *of* $L_{\kappa\kappa}$, *for* $\nu<\kappa$, $n\in\omega$, *such that*

$$\underset{\sim}{\mathfrak{A}} \models (\exists\vec{v}) \bigvee_{m\in\omega} \bigwedge_{n\in\omega} [(\forall\{u_\xi \mid \xi<\kappa_n\}) \bigvee_{\nu<\kappa} \varphi_{\nu,n}(\vec{v},\vec{u}) \wedge \bigwedge_{n\in\omega} \bigwedge_{\nu<\kappa} (\exists^{<\kappa_m}\vec{u})\varphi_{\nu,n}(\vec{v},\vec{u})],$$

*where* $(\exists^{<\kappa_m}\vec{u})\varphi(\vec{u})$, *for any formula* $\varphi$, *is an abbreviation for*

$$[\bigwedge_{\nu<\kappa_m} (\forall\{u_{\nu\xi} \mid \xi<\kappa_n\})\varphi(u_{\nu 0},\ldots,u_{\nu\xi},\ldots)] \longrightarrow$$

$$[\bigvee\{\bigwedge_{\xi<\kappa_n} u_{\nu_1\xi} = u_{\nu_2\xi} \mid \nu_1<\nu_2<\kappa_m\}].$$

The proof of the difficult part, (2) $\longrightarrow$ (3), utilizes the construction of a tree with $\kappa^\omega$ branches, much as Kueker's theorem for the countable case uses a binary tree construction, see his paper in this volume; recall that $\kappa^\omega = 2^\kappa$ if $\kappa$ is a strong limit cardinal of cofinality $\omega$.

The following corollary is the analogue of Kueker's result that a countable model which is $L_{\omega_1\omega}$-equivalent to an uncountable one has $2^\omega$ automorphisms.

3.5 COROLLARY. *Let* $\underset{\sim}{\mathfrak{A}}$ *be of strict power* $\kappa>\omega$. *If* $\underset{\sim}{\mathfrak{A}} \equiv_{\infty\kappa} \underset{\sim}{\mathfrak{B}}$, *where* $\underset{\sim}{\mathfrak{B}}$ *is* <u>*not*</u> *of strict power* $\kappa$, *then* $\underset{\sim}{\mathfrak{A}}$ *has* $2^\kappa$ *automorphisms.*

3.6 COROLLARY. *If* $\mathfrak{A}$ *is a standard model of power* $\kappa>\omega$, *and* $\mathfrak{A} \equiv_{\infty\kappa} \underset{\sim}{\mathfrak{A}}$, *where* $\underset{\sim}{\mathfrak{A}}$ *is a decomposition of* $\mathfrak{A}$ *of strict power* $\kappa$, *then* $\mathfrak{A}$ *has* $2^\kappa$ *automorphisms.*

PROOF: By the preceding corollary, the decomposition $\underset{\sim}{\mathfrak{A}}$ has $2^\kappa$ automorphisms. But an automorphism of the decomposition $\underset{\sim}{\mathfrak{A}}$ is an automorphism of the standard model $\mathfrak{A}$, and thus $\mathfrak{A}$ has $2^\kappa$ automorphisms. ⊣

At the opposite extreme are those chain models which are rigid (i.e., have only one automorphism). Among the interesting results is the following, proven by an application of Scott sentences:

3.7 THEOREM. *If* $\underset{\sim}{\mathfrak{A}}$ *is a chain model of strict power* $\kappa > \omega$ *which is rigid, then* $\underset{\sim}{\mathfrak{A}}$ *is definable up to isomorphism among* ***all*** *chain models by a sentence of* $L_{\kappa\kappa}$.

The proof of this theorem makes necessary use of infinite quantifiers, and thus it does not hold for $\kappa = \omega$. For example, let $\mathfrak{A}_n = \langle n,<\rangle$. Then $\underset{\sim}{\mathfrak{A}}$ is a chain model of strict power $\omega$ which is rigid, but there is no statement of $L_{\omega_1\omega}$ which defines $\underset{\sim}{\mathfrak{A}}$ up to isomorphism among chain models. For $\underset{\sim}{\mathfrak{A}}$ is $L_{\omega_1\omega}$-equivalent to its union $\langle\omega,<\rangle$, but cannot be chain isomorphic to $\langle\omega,<\rangle$ since $\langle\omega,<\rangle$ is not of strict power $\omega$.

We conclude with an example of a case where the analogy between countable models of $L_{\omega_1\omega}$ and chain models of $L_{\kappa\kappa}$ breaks down. If $\underset{\sim}{\mathfrak{A}}$ is a strict-power-$\omega$ decomposition of $\mathfrak{A}$, then every automorphism of $\mathfrak{A}$ is a chain automorphism of $\underset{\sim}{\mathfrak{A}}$, since clearly every automorphism of $\mathfrak{A}$ is bounded w.r.t. $\underset{\sim}{\mathfrak{A}}$. It had thus been conjectured by Karp that condition (1) of Theorem 3.3 was equivalent to

(*) Every automorphism of $\mathfrak{A}$ is bounded with respect to $\underset{\sim}{\mathfrak{A}}$, for some strict-power-$\kappa$ decomposition $\underset{\sim}{\mathfrak{A}}$ of $\mathfrak{A}$.

However, it is not difficult to show that if $\mathfrak{A}$, a standard model of power $\kappa$, has $\leq \kappa$ automorphisms, then (*) holds. Since, as we have just seen, (1) implies that $\mathfrak{A}$ has $2^\kappa$ automorphisms, it is clear that (*) $\Longrightarrow$ (1) is false.

## REFERENCES

1. C.C. Chang, *Some remarks on the model theory of infinitary languages*, in: The Syntax and Semantics of Infinitary Languages, Springer-Verlag, Berlin, 1968, 36-63.

2. E. Cunningham, *Chain Models for Infinite-Quantifier Languages*, Ph.D. Thesis, University of Maryland, 1974.

3. J. Green, *Consistency Properties for Uncountable Finite-Quantifier Languages*, Ph.D. Thesis, University of Maryland, 1972.

4. C. Karp, *Infinite-quantifier languages and* $\omega$*-chains of models*, in: Proceedings of the Tarski Symposium, American Mathematical Society, Providence, 1974, 225-232.

5. C. Karp, *Interpreting formal languages in directed systems of structures*, J. Symbolic Logic 29(1964), abstract, 155.

6. C. Karp, Languages with Expressions of Infinite Length, North-Holland, Amsterdam, 1964.

7. H.J. Keisler, Model Theory for Infinitary Logic, North-Holland, Amsterdam, 1971.

8. D.W. Kueker, *Definability, automorphisms, and infinitary languages*, in: The Syntax and Semantics of Infinitary Languages, Springer-Verlag, Berlin, 1968, 152-165.

9. M. Makkai, *Generalizing Vaught sentences from* $\omega$ *to strong cofinality* $\omega$, to appear.

PART D

# ON A FINITENESS CONDITION FOR INFINITARY LANGUAGES

BY

JOHN GREGORY

# CONTENTS
## PART D

# INTRODUCTION

The compactness theorem for finitary languages is:

For all collections $\Gamma$ of sentences, $\Gamma$ has a model if each finite subset of $\Gamma$ has a model.

In what sense can the compactness theorem be generalized to infinitary languages? In [14; §13], Kreisel formulated a "generalized finiteness theorem," which we call GFT. He also asked whether GFT was actually a theorem for every admissible set. It is not. This fact is our Theorem 1.1.

In subsection 0.5, we state GFT and a weaker WFT. In the last part of this section, we list some related facts given in [2], [15], and [3]; we will not use these. We also give a quick example of a failure of WFT when individuals are allowed. (Unless otherwise specified, we assume the axioms ZFC of Zermelo-Fraenkel set theory with the axiom of choice, with the axiom of regularity, and without any individuals.) In §1, we prove Theorem 1.1 modulo Lemma 1.2 proved in §5. In §2, we prove a simplified version of Lemma 1.2 and Theorem 1.1 when individuals are allowed. In §3, we briefly describe the Boolean version of forcing used here. In §4, we construct a number of Boolean extensions of a given admissible set (using what is essentially Cohen's original forcing conditions [6]). In §5, Lemma 1.2 is proved by coherently reducing the Boolean extensions of §4 to transitive sets, assuming the given admissible set is countable. In §6 of our dissertation, omitted in this volume, we proved some consistency results. For example, the following are consistent with ZFC:

(6.8) For every admissible $\alpha > \omega$ there exists an admissible set $X$ for which WFT fails and $o(X) = \alpha$;

(6.10, 6.11, 6.13) There are admissible sets $X$ for which WFT fails and each $X$-a.i.d. set is an element of $X$.

We assume that the reader understands the equivalent of [9] and

[22].

We will frequently use the terminology of the primitive recursive (set) functions and relations of [9]. This will be primarily a device for stating that a function or relation is defined in a sufficiently effective manner.

## SECTION 0

### PRELIMINARIES

0.1. The infinitary finite-quantifier language $L$ of [2] allows infinite conjunctions but not infinite quantifiers. The usual notions of syntax are primitive recursive without parameters. For technical convenience, we now list the basic objects of a specific version of the language $L$. (These objects are considered to be sets. For some of the set theory notation, see the end of this section; in particular, $\langle x,y\rangle$ is the ordered pair.)

Terms: Constants: $\langle 0,x\rangle$ for each set $x$

Variables: $\langle 1,x\rangle$ for each set $x$

Relation symbols: for each $n \in \omega$, $n$-placed relation symbols:

$\langle 2,x,n\rangle$ for each set $x$.

The collection of infinitary formulas is inductively defined. A set is a formula iff it is one of the following:

Atomic formula $\underline{R}(t_1,\ldots,t_n)$: $\langle \underline{R},t_1,\ldots,t_n\rangle$ for some $n$, some $n$-placed relation symbol $\underline{R}$, some terms $t_1,\ldots,t_n$

Negation $\neg\varphi$: $\langle 3,\varphi\rangle$ for some formula $\varphi$.

Conjunction $\Lambda x$: $\langle 4,x\rangle$ for some set $x$ of formulas.

Universal quantification $\forall v\varphi$: $\langle 5,v,\varphi\rangle$ for some variable $v$ and formula $\varphi$.

Then $\Lambda x$ is interpreted as "for all $\varphi$ in $x$, $\varphi$." Define disjunction $\vee x = \neg\Lambda\{\neg\varphi \mid \varphi\in x\}$; then $\vee x$ is interpreted as "for some $\varphi$ in $x$, $\varphi$." Define $\varphi \,\&\, \psi$, $\varphi \vee \psi$, $\varphi\rightarrow\psi$, $\varphi\leftrightarrow\psi$, $\exists v\varphi$ to be $\Lambda\{\varphi,\psi\}$,

$\vee\{\varphi,\psi\}$, $\neg\varphi \vee \psi$, $(\varphi \to \psi) \& (\psi \to \varphi)$, $\neg\forall \vee \neg\varphi$, respectively. The equality symbol $\equiv$ we identify with $<2,0,2>$; unless specified otherwise, it is interpreted by the equality relation $=$. The symbol $\varepsilon$ we identify with $<2,1,2>$.

For our later examples of collections $\Gamma$ of infinitary formulas, to each $x$ we assign a constant $\underline{x} = <0,<0,x>>$. We will assume that all other constants, variables, and relation symbols of $\Gamma$ are hereditarily finite sets.

We use the usual terminology for syntax. The prefix "$L$-" is used to emphasize that infinite conjunctions are allowed.

The ordinary, i.e., finitary (first order), language will be identified with the sublanguage of infinitary formulas $\varphi$ such that: all variables and relation symbols occurring in $\varphi$ are hereditarily finite sets (we will only need countably many of these); all conjunctions $\wedge x$ occurring in $\varphi$ are such that $x$ has only one or two elements, i.e., $\wedge x = \chi \& \psi$ for some $\chi$ and $\psi$.

We use an auxiliary language $L^+$ with names. To each set $x$ we assign a name $\overline{x} = <9,x>$. Each name is a constant of the language $L^+$; the other constants of $L^+$ are those of $L$. Then variables, relation symbols, and the logical operations of $L^+$ are the same as those of $L$. We also have the finitary language with names.

Each name $\overline{x}$ is understood to be interpreted by $x$. If $\overline{x}_1,\ldots,\overline{x}_n$ are the names occurring in formula $\varphi$, then $x_1,\ldots,x_n$ are called the parameters of $\varphi$. In writing formulas, we generally omit the overlines; formulas $\varphi(\overline{x}_1,\ldots,\overline{x}_n)$ is written $\varphi(x_1,\ldots,x_n)$. When we discuss a formula $\varphi$, $\vec{x}$ is representative of parameters that might occur in $\varphi$; $\vec{v}$ is a representative of corresponding variables. We act as if $\vec{x}$ were a single parameter. For a set (or definable class) $U$, a $U$-formula is a $L^+$ formula such that all its parameters are elements of $U$. Similarly, we have $U$-sentences.

We also use $\forall$, $\exists$, etc., for abbreviations of English quantifications, etc. The context usually determines the use. In particular, $\epsilon$ never occurs in a $L^+$ formula.

A structure $\mathcal{U}$ = <U;G;H> consists of the following:

a nonempty set U;

a function G from a set of relation symbols to relations on U, where n-placed relation symbols are mapped by G to n-ary relations; and

a function H from a set of constants of $L$ into U.

As mentioned above, G(≡) is generally assumed to be the equality relation restricted to U. If $\{\equiv,\underline{R}_1,\ldots,\underline{R}_m\}$ is the domain of G and if H is empty, we ambiguously write $\langle U,G(\underline{R}_1),\ldots,G(\underline{R}_m)\rangle$ for $\mathcal{U}$. Frequently, we notationally identify the structure with its universe U.

The satisfaction relation ⊨ is defined by the usual induction. We list the determining factors.

<U;G;H> ⊨ φ iff:

<U;G;H> is a structure $\mathcal{U}$;

φ is a U-sentence;

the relation symbols of φ are in the domain of G;

the $L$ constants of φ are in the domain of H;

if φ is some atomic $\underline{R}(t_1,\ldots,t_n)$, then $\langle F(t_1),\ldots,F(t_n)\rangle \in G(\underline{R})$ for the assigning function $F = H \cup \{\langle\overline{x},x\rangle \mid x\in U\}$;

if φ is some ¬ψ, then not $\mathcal{U} \models \psi$;

if φ is some ⋀x, then $\forall\psi \in x(\mathcal{U} \models \psi)$; and if $\varphi$ is some $\forall v\psi(v)$, then $\forall y \in U\,(U \models \psi(y))$.

A structure $U$ is a model of a collection $\Gamma$ of $L$-sentences iff $\forall\varphi \in \Gamma\,(U \models \varphi)$.

For a function $f$ on variables and for formula $\varphi$, define $\varphi[f]$ to be the result of replacing the free occurrences of $v$ in $\varphi$ by $\overline{f(v)}$ for each $v$ in the domain of $f$. Thus, if $f(v_i) = x_i$ for each $i \leq n$, then $\varphi(v_0,\ldots,v_n)[f]$ is $\varphi(x_0,\ldots,x_n)$.

0.2. The implicit definability notions below are generalizations of finiteness, recursiveness, and recursive enumerability. That is, if $X$ is the set $\omega$ of natural numbers, then the X-a.i.d. sets are the finite subsets of $X$, the X-i.i.d. sets are the recursive sets, and the X-s.i.i.d. sets are the recursively enumerable sets.

Given a (nonempty) transitive set $X$, the usual interpretation of the binary relation symbol $\varepsilon$ is $\epsilon_X = \{\langle x,y\rangle \mid x\epsilon y \;\&\; y\epsilon X\}$, which is the restriction to $X$ of the membership relation $\epsilon$. The subscript will be omitted. For any binary relation $E$, we have the following definition.

<u>Definition.</u> A structure $\langle Y,E\rangle$ is an end extension of a transitive set $X$ iff $Y \supseteq X$ and $\{y \mid y\,E\,x\} = x$ for all $x \in X$. Example: any transitive set $Y \supseteq X$ with $E = \epsilon$ (restricted to $Y$).

<u>Definition.</u> X-a.i.d. (absolutely implicitly definable). A set $R$ is X-a.i. defined by $\varphi$ iff

(0) $\varphi$ is a finitary X-sentence in some relation symbols $\varepsilon$, $=$, $\underline{R}$, $\underline{S}_1,\ldots,\underline{S}_n$; $R \subseteq X$

(1) there are relations $S_1,\ldots,S_n$ such that $\langle X,\epsilon,R,S_1,\ldots,S_n\rangle \models \varphi$

(2) for all end extensions $\langle Y,E\rangle$ of $X$ and for all $R'$, $S_1',\ldots,S_n'$, if $\langle Y,E,R',S_1',\ldots,S_n'\rangle \models \varphi$, then $R' = R$.

Then $\varphi$ is called a X-a.i. definition iff it X-a.i. defines some set R. A set R is called X-a.i.d. (absolutely implicitly definable, or X-finite) iff it is X-a.i. defined by some $\varphi$. Note that each element of X is X-a.i.d. (see below). Note that $\varphi$ uniquely determines R if the principal relation symbol $\underline{R}$ is specified. Condition (2) is a persistence requriement; it attempts to give a predicative character to the definition.

For the definition of X-i.i.d. (invariantly implicitly definable, or X-rec), replace "$R' = R$" in clause (2) by "$R' \cap X = R$." For the definition of X-s.i.i.d. (semi-invariantly implicitly definable, or X-re), replace "$R' = R$" by "$R' \supseteq R$." All X-gen$\Sigma_1$ sets are X-s.i.i.d. (X-gen$\Sigma_1$ is defined below). All X-gen$\Delta_1$ sets are X-i.i.d.

Example. Each element x of X is X-a.i.d. For let $\varphi$ be the X-sentence $\forall v(\underline{R}(v) \leftrightarrow v \epsilon x)$. Considering x as a unary relation R, $\langle X, \epsilon, x\rangle \models \varphi$. If $\langle Y, E\rangle$ is an end extension of X and if $\langle Y, E, R'\rangle \models \varphi$, then $\forall y \in Y(y \epsilon R' \leftrightarrow y E x)$; then $R' = \{y \mid y E x\} = x$.

Example. The transitive closure of a X-a.i.d. set is also X-a.i.d. For let $\psi$ be a X-a.i. definition of $R_1$ and suppose $\underline{R}_1$ is the principal relation symbol. Let $\underline{R}$ and $\underline{S}$ be unary and binary relation symbols, respectively, not occurring in $\psi$. Let u, v, w be distinct variables. Let $\varphi$ be

$$\psi \ \& \ \forall uv(\underline{S}(u,v) \longleftrightarrow \exists w \ \epsilon u(w \equiv v \vee \underline{S}(w,v)))$$

$$\& \ \forall v(\underline{R}(v) \longleftrightarrow \exists u(\underline{R}_1(u) \ \& \ (u = v \vee \underline{S}(u,v)))).$$

Then $\varphi$ is a X-a.i. definition of $TC(R_1)$, where $\underline{R}$ is the principal relation symbol corresponding to $TC(R_1)$.

We hereafter assume that the principal relation symbol is specified to be, say, $\langle 2,0,1\rangle$. We write $\underline{S}$ for $\underline{S}_1,\ldots,\underline{S}_n$.

Note that the persistence property (2) for a.i. definability implies that: if $X \subseteq X'$, if $\varphi$ is a X-a.i. definition of R, and if $\varphi$ is a X'-a.i. definition of R', then R = R'. A similar remark holds for s.i.i. definability (that is, if $X \subseteq X'$, if $\varphi$ is a X-s.i.i. definition of R, and if $\varphi$ is a X'-s.i.i. definition of R', then $R \subseteq R'$).

0.3. We write $L\varepsilon$ for the set of finitary formulas of the (pure) language of set theory; i.e., $L\varepsilon$ is the set of formulas in the relation symbol $\varepsilon$ (no other relation symbols except, when convenient, $\equiv$; no names or constants). The sets $\Sigma_0$, $\Pi_0$, gen$\Sigma_n$, and gen$\Pi_n$, of $L\varepsilon$ formulas, have the usual definitions, as follows. (Read "gen" as "generalized.") A quantifier $\forall u$ is restricted iff it occurs in the context $\forall u(u\varepsilon v \rightarrow\ )$ for some variable v; we write $\forall u \varepsilon v(\ )$ for $\forall u(u\varepsilon v\rightarrow\ )$. Similarly, we have restricted quantifiers $\exists u(u\varepsilon v\ \&\ )$, written $\exists u \varepsilon v(\ )$. Then $\Sigma_0$, $\Pi_0$, gen$\Sigma_0$, and gen$\Pi_0$ are each the set of $L\varepsilon$ formulas $\varphi$ such that all quantifiers occur restricted in $\varphi$. Then gen$\Sigma_{n+1}$ is the set of $L\varepsilon$ formulas $\varphi$ built up from gen$\Sigma_n$ and gen$\Pi_n$ formulas by using &, $\vee$, $\forall u \varepsilon v$, $\exists u \varepsilon v$, and $\exists$. Similarly, gen$\Pi_{n+1}$ is the set of $L\varepsilon$ formulas built up from gen$\Sigma_n$ and gen$\Pi_n$ formulas by using &, $\vee$, $\forall u \varepsilon v$, $\exists u \varepsilon v$, and $\forall$. If all the unrestricted quantifiers of a gen$\Sigma_n$ formula occur in front, then it is called a $\Sigma_n$ formula (similarly for $\Pi_n$).

For a set $\Phi$ (e.g., $\Sigma_0$) of $L\varepsilon$ formulas, a formula will be called a $\Phi^+$ formula iff it is the result $\psi(\overline{x}_1,\ldots,\overline{x}_n)$ (usually written $\psi(x_1,\ldots,x_n)$) of substituting names of some $x_1,\ldots,x_n$ into some $\Phi$ formula $\psi(v_1,\ldots,v_n)$. Thus, a $\Phi^+$ formula may contain parameters.

When X is a transitive set and $\Phi$ is a set of $L\varepsilon$ formulas, then a n-ary relation R is X - $\Phi$ iff there is some $\Phi^+$ X-formula $\varphi(v_1,\ldots,v_n)$ which X-defines R, i.e.,

$$\forall x_1 \ldots x_n (\langle x_1,\ldots,x_n\rangle \in R \longleftrightarrow \langle X,\in\rangle \models \varphi(x_1,\ldots,x_n)).$$

The relation is $X$-$\Phi$ without parameters iff it has a $X$-defining $\Phi$ formula.

A relation is $X$-gen$\Delta_n$ iff it is both $X$-gen$\Sigma_n$ and $X$-gen$\Pi_n$; similarly, a $X$-$\Delta_n$ relation is both $X$-$\Sigma_n$ and $X$-$\Pi_n$.

The $\Sigma_0$ formulas $\varphi(\vec{v})$ have an absoluteness property: if $X$ is transitive, $\vec{x} \in X$, and $\langle Y,E\rangle$ is an end extension of $X$, then

$$\langle X,\in\rangle \models \varphi(\vec{x}) \quad \text{iff} \quad \langle Y,E\rangle \models \varphi(\vec{x}).$$

The gen$\Sigma_1$ formulas $\varphi(\vec{v})$ have an important persistence property: if $X$ is transitive, $\vec{x} \in X$, and

$$\langle X,\in\rangle \models \varphi(\vec{x}),$$

then, for every end extension $\langle Y,E\rangle$ of $X$,

$$\langle Y,E\rangle \models \varphi(\vec{x}).$$

Thus, as mentioned earlier, $X$-gen$\Sigma_1$ sets are $X$-s.i.i.d. For if $\vec{x} \in X$ and $\varphi(v,\vec{v})$ is gen$\Sigma_1$, then $\forall v(\underline{R}(v) \leftrightarrow \varphi(v,\vec{x}))$ is a $X$-s.i.i. definition, by the persistence property just mentioned.

0.4. An admissible set $X$ is a transitive set with the following axiomatizable properties:

$x,y \in X$ implies $\{x,y\} \in X$ and $\bigcup x \in X$

$\Sigma_0$-separation principle: if $R$ is $X$-$\Sigma_0$ and $x \in X$, then $x \cap R \in X$.

$\Sigma_0$-reflection principle: if binary $R$ is $X$-$\Sigma_0$ and $x \in X$, then

$$\forall y \in x\ \exists z \in X\ R(y,z) \longrightarrow \exists w \in X\ \forall y \in x\ \exists z \in w\ R(y,z).$$

The corresponding axioms are the universal closures of the following formulas (where $v$, $v_0$, $v_1$, $v_2$ are distinct variables):

pair: $\exists v_2 \forall v_1 (v_1 \varepsilon v_2 \leftrightarrow v_1 = v \vee v_1 = v_0)$

union: $\exists v_2 \forall v_1 (v_1 \varepsilon v_2 \leftrightarrow \exists v_0 \varepsilon v (v_1 \varepsilon v_0))$

$\Sigma_0$-separation schema: $\exists v_2 \forall v_1 (v_1 \varepsilon v_2 \leftrightarrow v_1 \varepsilon v \,\&\, \varphi)$ for each $\Sigma_0$ formula $\varphi$ such that $v_2$ is not free in $\varphi$

$\Sigma_0$-reflection schema: $\forall v_0 \varepsilon v \exists v_1 \varphi \rightarrow \exists v_2 \forall v_0 \varepsilon v \exists v_1 \varepsilon v_2 \varphi$ for each $\Sigma_0$ formula $\varphi$ such that $v_2$ is not free in $\varphi$.

(Actually, there is a finite subcollection of these axioms such that the admissible sets are precisely those transitive sets which are models of the finite subcollection ($\in$ interprets $\varepsilon$).)

For the remainder of this section, a set denoted by $X$ is understood to be an admissible set.

The above definition of admissible sets is equivalent to those given in [2] and [15]. Each gen$\Sigma_1$ formula is $X$-equivalent to a $\Sigma_1$ formula of form $\exists v\varphi$ where $\varphi$ is $\Sigma_0$. Thus, each $X$-gen$\Sigma_1$ relation is a $X$-$\Sigma_1$ relation; each $X$-gen$\Delta_1$ relation is a $X$-$\Delta_1$ relation. Further, $X$ satisfies the $\Delta_1$-separation principle:

$$\text{if } R \text{ is } X\text{-}\Delta_1 \text{ and } x \in X, \text{ then } x \cap R \in X.$$

Also, $X$ satisfies the $\Sigma_1$-reflection principle. (In this paper, if we have some notion of $X$-$\mathcal{R}$ relations on $X$, then the $\mathcal{R}$-reflection principle is the result of changing "$X$-$\Sigma_0$" to "$X$-$\mathcal{R}$" in the statement of the $\Sigma_0$-reflection principle. Similarly, we can define the $\mathcal{R}$-separation principle. The notion of $\Sigma_1$-reflection is particularly important because of the persistence property for $\Sigma_1$ formulas. The axioms of pair and union imply the equivalence of the $\Sigma_0$-, $\Sigma_1$-, and gen$\Sigma_1$-reflection principles for a given transitive set.)

A theory of recursion for admissible sets has been developed by Platek in [20].

The restrictions to $X$ of the primitive recursive (set) functions of [9] are totally X-defined by their gen$\Sigma_1$ definitions given in [9]; these restrictions are X-$\Delta_1$ (here, a function is considered to be a relation). We make frequent use of this fact; this is part of what we meant earlier by "sufficiently effective."

We will say that a n-place function F is primitive recursive with parameters $\vec{x}$ iff there is some primitive recursive function G such that $G(\vec{x},x_1,\ldots,x_n) = F(x_1,\ldots,x_n)$ for all $x_1,\ldots,x_n$. Thus, if a function is primitive recursive with parameters in X, then its restriction to X is X-$\Delta_1$ and total. There is a similar definition and remark for relations.

The notion of Prim-closed was defined in [9]. A set Y is Prim-closed iff every n-place primitive recursive function maps $Y^{(n)}$ into Y. The equivalent definition [9; 4.2] of admissible sets says that a nonempty set is admissible iff it is transitive, Prim-closed, and satisfies the $\Sigma_1$-reflection principle.

0.5. For the infinitary finite-quantifier language $L$ and for admissible set X, we now give (a equivalent version of) Kreisel's GFT.

The "generalized finiteness theorem," GFT, is:

> For all X-s.i.i.d. collections $\Gamma$ of $L$-sentences, $\Gamma$ has a model if each X-a.i.d. subset of $\Gamma$ has a model.

Compare this with the (true) compactness theorem for finitary language. The reader is referred to [14; §13] for the considerations that led to the formulation of GFT.

This GFT implies the following property, WFT:

> For all X-$\Delta_1$ collections $\Gamma$ of $L$-sentences, has a model if each X-a.i.d. subset of has a model.

It can be shown that the result of replacing "$X\text{-}\Delta_1$" in WFT by either "X-i.i.d." or "$X\text{-}\Sigma_1$" is still equivalent to WFT. As in Theorem 2.8 of [3], one can show that WFT is equivalent to:

> For all X-s.i.i.d. collections $\Gamma$ of $L$-sentences, $\Gamma$ has a model if each X-a.i.d. set R is such that $R \cap \Gamma$ has a model.

Note that we allow parameters in all our notions of $X\text{-}\Phi$ sets. Here, most $\Gamma$ will be $X\text{-}\Delta_1$ without parameters.

There are other generalizations of the finitary compactness theorem. If the notion of X-a.i.d. is replaced by the notion of being an element of X, then we have the notion of s.i.i.d.-compactness:

> For all X-s.i.i.d. collections $\Gamma$ of $L$-sentences, $\Gamma$ has a model if:
>
> $$\forall \Gamma_0 \in X(\Gamma_0 \subseteq \Gamma \longrightarrow \Gamma_0 \text{ has a model}).$$

Then s.i.i.d.-compactness implies GFT. Barwise (in [2], [3], [4]) has investigated the similar notion of $\Sigma_1$-compactness (replace "X-s.i.i.d." by "$X\text{-}\Sigma_1$" in the definition of s.i.i.d.-compactness).

0.6. In [2], Barwise proved for every <u>countable</u> admissible X:

1) X is $\Sigma_1$-compact.

2) If $\Gamma$ is a $X\text{-}\Sigma_1$ collection of $L$-sentences, then $\{x \in X \mid \text{every model of } \Gamma \text{ is a model of } x\}$ is $X\text{-}\Sigma_1$, i.e., then the set of logical consequences in X of $\Gamma$ is $X\text{-}\Sigma_1$.

3) Each X-a.i.d. set is an element of X.

In [15], Kunen proved for every admissible X:

4) The set of logical consequences in X of $\Gamma$ is X-s.i.i.d.

whenever $\Gamma$ is a X-s.i.i.d. collection of $L$-sentences. The proof gives a good example of a s.i.i. definition.

5) If R is X-s.i.i.d., then there is an $X\text{-}\Delta_1$ collection $\Gamma$ of $L$-sentences and there is a $X\text{-}\Delta_1$ function F such that, for all $x \in X$, $F(x)$ is a formula, and $x \in R$ iff $F(x)$ is a logical consequence of $\Gamma$.

6) By (2) and (5), X-s.i.i.d. is $X\text{-}\Sigma_1$ whenever X is countable.

7) Each X-s.i.i.d. set is $X\text{-}\Pi^1_1$ (and some $X\text{-}\Pi^1_1$ sets are not $X\text{-}\Sigma_1$ and hence not X-s.i.i.d. when X is countable).

8) Suppose X is X-a.i.d. Then X is called self-definable. All X-i.i.d. sets are X-a.i.d. The X-s.i.i.d. sets are the $X\text{-}\Pi^1_1$ sets. The X-a.i.d. sets are the $X\text{-}\Delta^1_1$ sets. ($H_{\aleph^+}$ is known to be self-definable.)

By (1), (3), and (6), s.i.i.d.-compactness and GFT hold for any countable admissible set.

When X is self-definable, any $X\text{-}\Delta_1$ collection $\Gamma$ of $L$-sentences is already X-a.i.d. Thus, WFT is trivial for self-definable X. We do not know whether GFT holds for self-definable X.

In an earlier version of [15], Kunen mentioned the following condition on X:

There is some finitary X-sentence $\psi$ (we could allow $\psi$ to be $\Sigma^1_1$) such that:

$$\langle X, \in\rangle \models \psi; \quad \text{and}$$

For all end extensions $\langle Y,E\rangle$ of X such that $X \neq Y$ and $\langle Y,E\rangle \models \psi$:

$$\exists y \in Y \forall x \in X (x \, E \, y).$$

He showed that, if X satisfies this condition and if X-s.i.i.d.

compactness fails, then $X$ is self-definable. Therefore, WFT holds for $X$ whenever the condition holds for $X$. The condition holds for $X$ built up in a sufficiently neat linear manner (a good collection of such $X$ is given by [4; 5.2.2]), for example, any admissible $V(\alpha)$ or $L(\alpha)$ (where $V(\alpha)$ is the set of all sets of rank less than $\alpha$; $L(\alpha)$ is a segment of the constructible hierarchy of [9]).

In looking for a failure of GFT, one had to look at uncountable admissible sets. The usual uncountable admissible sets are covered in the last two paragraphs. A proof of the failure of GFT for some such $X$ (if there is a proof) would seem to involve finding a $X$-$\Pi_1^1$ collection $\Gamma$ of $L$-sentences such that each $X$-$\Delta_1^1$ subset of $\Gamma$ has a model, but $\Gamma$ does not. It would be interesting to know if such a $\Gamma$ exists for $H_{\aleph_1}$. However, once one looks for uncountable sets not satisfying the condition of the last paragraph, one finds a plentiful supply of admissible sets failing GFT. The involved $\Gamma$'s are the simplest imaginable. Our reason for introducing WFT is to point this out.

0.7. The following observation has a straightforward modification that shows that it is at least consistent with ZFC that WFT can fail.

The failure of WFT is easy to show when individuals are allowed both in set theory (as in [26]) and in admissible sets. Suppose given two uncountable sets $I$ and $J$ of individuals such that $I \subseteq J$, $I \neq J$, and $I \simeq J$. Form the set $H_{\aleph_1}^I$ of hereditarily countable sets built up from elements of $I$. I.e., $H_{\aleph_1}^I$ is the set of all $x$ such that: $TC(x) \preceq \omega$; all individuals contained in $TC(x)$ are elements of $I$. Similarly, form $H_{\aleph_1}^J$. By the axiom of choice, $H_{\aleph_1}^I$ is admissible.

We now show that every $H_{\aleph_1}^I$-a.i.d. set is built up from countably many individuals because every element of $H_{\aleph_1}^I$ is. Suppose $\langle H_{\aleph_1}^I, \in, R, S\rangle \models \varphi(\vec{x})$ and $\varphi$ is a $H_{\aleph_1}^I$-a.i. definition of $R$. Suppose that TC(R) contains uncountably many individuals. Find individual $i \in TC(R) - TC(x)$. Find 1-1 onto $g: I \longrightarrow J$ such that $g \mid TC(x)$ is identity and $g(i) \in J - I$. Extend $g$ to all sets built up from elements of $I$ in the usual manner, i.e., $gx = g''x$ for non-individuals $x$). Then $\langle gH^I_{\aleph_1}, g\in, gR, gS\rangle \models \varphi(gx)$. Thus, $\langle H^J_{\aleph_1}, \in, gR, gS\rangle \models \varphi(\vec{x})$. But $H^J_{\aleph_1}$ is an end extension of $H^I_{\aleph_1}$. Also, $g(i) \in TC(gR) - TC(R)$. Thus, $gR \neq R$, which contradicts the persistence property.

Thus, $H^I_{\aleph_1}$ contains uncountably many individuals, but each $H^I_{\aleph_1}$-a.i.d. set contains only countably many individuals. It follows that WFT fails for $H^I_{\aleph_1}$. For let $\Gamma$ consist of the sentences

1) $\underline{a} \neq \underline{b}$ for each $a \neq b$ of $H^I_{\aleph_1}$

2) $\forall uvw(\underline{R}(u,v) \,\&\, \underline{R}(u,w) \longrightarrow v \equiv w)$

3) $\bigvee_{\alpha \in \omega} \underline{R}(\underline{\alpha},\underline{a})$ for each $a \in H^I_{\aleph_1}$ such that $\forall x \in a(x \notin a)$, i.e., $a \in I$ or $a = 0$.

Then $\Gamma$ has no model: Suppose $\Gamma$ had a model $\langle U,R,F\rangle$, where $R$ interprets $\underline{R}$ and $F(\underline{a})$ interprets $\underline{a}$. By sentences (1), $F(\underline{a}) \neq F(\underline{b})$ for distinct $a,b \in H^I_{\aleph_1} \supseteq$ . By sentence (2), $R$ is a function. By sentences (3), countable $R''\{F(\underline{\alpha}) \mid \alpha \in \omega\} \supseteq \{F(\underline{a}) \mid a \in I\} \simeq I$. Since $I$ is uncountable, this is impossible.

But each $H^I_{\aleph_1}$-a.i.d. subcollection $\Gamma_0$ of has some model: Let $F(\underline{a}) = a$ for each $a \in H^I_{\aleph_1}$. Let $R$ be a partial function from $\omega$ onto $\{a \mid \bigvee_{\alpha \in \omega} \underline{R}(\underline{\alpha},\underline{a})$ is contained in $_0\} \subseteq (I \cup \{0\}) \cap TC(\Gamma_0)$; this is possible since the last set was countable. Then $\langle H^I_{\aleph_1}, R, F\rangle$ is a model of $\Gamma_0$.

0.8. Even though GFT does fail for some uncountable admissible sets, it holds for many others. For the following examples, GFT holds in the stronger form of s.i.i.d.-compactness.

Barwise [3] and Karp [12] proved that GFT holds for many

admissible $V(\alpha)$. It holds for all $V(\alpha)$ such that $\alpha$ is cofinal with $\omega$ and $V(\alpha)$ is P-admissible (i.e., allow $\Sigma_0$ formulas to contain a relation symbol to be interpreted by $P$ in the definition of admissible set), where $P$ is the graph of the power set function restricted to $V(\alpha)$. GFT also holds for some $V(\alpha)$ such that $\alpha$ is not cofinal with $\omega$ [3].

It follows directly from the persistence property of s.i.i.d. that: for every infinite cardinal $\aleph$ and for every $x \in H_{\aleph^+}$, there are admissible $X$ such that $x \in X \simeq \aleph$ and GFT holds for $X$. (Barwise points out that this follows from Theorem 5.3 and (the proof of) Theorem 4.7 of his [4], using a second order constructible hierarchy $L^1_\alpha(x)$.) For this, we give a proof related to the proof of Lemma 6.9 below.

PROOF: Let regular uncountable cardinal $\kappa$ be given (e.g., $\aleph^+$). Define $M_\beta \in H_\kappa$ by induction on $\beta < \kappa$ as follows.

$M_0$ is some given admissible set contained in $H_\kappa$.

$M_\beta = \bigcup_{\alpha<\beta} M_\alpha \in H_\kappa$ for limit $\beta \neq 0$.

$M_{\alpha+1}$ is defined by the next two paragraphs.

For each finitary $M_\alpha$-sentence $\varphi$ of the definition of a.i.d. (hence, of s.i.i.d.):
If $M_\alpha$ is a subset of some transitive $M \prec \kappa$ such that

$\varphi$ is a $M$-s.i.i. definition of a collection $\Gamma$ of $L$-sentences such that $\Gamma$ has no model,

then define $G(\varphi) = \{M,\Gamma\}$ for a choice of some such $M$.

Then function $G \in H_\kappa$: $\mathrm{dom}(G) \in H_\kappa$ and $\mathrm{rng}(G) \subseteq H_\kappa$; by regularity of $\kappa$, $G \in H_\kappa$. Let $M_{\alpha+1}$ be a transitive elementary submodel of $H_\kappa$ such that $G \in M_{\alpha+1} \in H_\kappa$.

Consider any limit $\beta \neq 0$ where $\beta < \kappa$. Then $M_\beta \prec \kappa$ and $M_\beta$

is a transitive elementary submodel of $H_\kappa$. It is admissible. We will show that $M_\beta$ is s.i.i.d.-compact.

Suppose $\varphi$ is a $M_\beta$-s.i.i. definition of a collection $\Gamma'$ of $L$-sentences such that $\Gamma'$ has no model; we show that there is $\Gamma \in M_\beta$ where $\Gamma \subseteq \Gamma'$ and $\Gamma$ has no model. Now $\varphi$ is a $M_\alpha$-sentence for some $\alpha < \beta$. At the $\alpha$-th stage, $G(\varphi)$ has some value $\{M,\Gamma\}$, since $M_\beta$ satisfies the "if" clause in the definition of G. Then $M \subseteq M_\beta$. By persistence, $\Gamma' \supseteq \Gamma$. Further, $\Gamma$ has no model and $\Gamma \in M_\beta$. QED

(For limit $\beta \neq 0$, the above $M_\beta$ has the s.i.i.d.-separation property.)

0.9. We now list some of the definitions and known facts used later. Only (1), (2) and (3) are used for Theorem 1.1.

(1) The set $\aleph_1$ of all countable ordinals has the ordinal ordering $<$, which is the same as $\in$. This induces the order topology on $\aleph_1$; then A is a closed subset of $\aleph_1$ iff, for every $S \subseteq A$, the least upper bound $\cup S$ is either $\aleph_1$ or an element of A. Since $\aleph_1$ is regular, a subset A of $\aleph_1$ is uncountable iff A is cofinal with $\aleph_1$.

Let $\mathcal{F}$ be the set of all uncountable closed subsets of $\aleph_1$. The following is a special case of [1;§7, Satz 4].

If countable $C \subseteq \mathcal{F}$, then $\cap C \in \mathcal{F}$.

(2) Let $\mathcal{U}$ and $\mathcal{U}'$ be structures with universes U and U', respectively. Then $\mathcal{U}'$ is an elementary extension of $\mathcal{U}$ (i.e., $\mathcal{U}$ is an elementary submodel of $\mathcal{U}'$) iff:

$U \subseteq U'$ and, for every U-sentence $\varphi$,
$\mathcal{U} \models \varphi$ iff $\mathcal{U}' \models \varphi$.

A directed system of elementary extensions is a set $S$ of structures such that for every $\mathfrak{U}, \mathfrak{U}' \in S$, there is $\mathfrak{U}'' \in S$ where $\mathfrak{U}''$ is an elementary extension of both $\mathfrak{U}$ and $\mathfrak{U}'$.

The following is a known fact of model theory: The union of a directed system $S$ of elementary extensions is an elementary extension of each member of $S$.

(3) For convenience, sometimes we will call a set $f$ (or a definable class $f$) a (partial) primitive recursive function if $f$ is a restriction of a primitive recursive function $F$ to a primitive recursive domain $D$:

$$\langle \vec{x}, y\rangle \in f \quad \text{iff both} \quad \vec{x} \in D \text{ and } F(\vec{x}) = y.$$

For some purposes we can identify $f$ with $F$.

(4) The following (a), (b) are from [18].

(a) For every gen$\Sigma_1$ formula $\psi(\vec{v})$ there is a $\Sigma_0$ formula $\varphi(\vec{v}, v)$ such that for all $\vec{x}$

$$\langle V, \in\rangle \models \psi(\vec{x}) \qquad \text{iff} \qquad \langle V, \in\rangle \models \exists v\varphi(\vec{x}, v).$$

($V$ is the class of all sets.)

(b) For every $\Sigma_0$ formula $\varphi(\vec{v}, v)$, if $\vec{x} \in H_{\aleph_1}$, then

$$\langle V, \in\rangle \models \exists v\varphi(\vec{x}, v) \qquad \text{iff} \qquad \langle H_{\aleph_1}, \in\rangle \models \exists v\varphi(\vec{x}, v).$$

(c) We say that $x$ is contructible from $s$ iff

$$y \in L_t = \bigcup_{\beta} L_t(\beta)$$

where $t = TC(\{s\})$ and $L_t$ was defined in [9; 2.10]. There is a gen$\Sigma_1$ formula $\psi(v_0, v_1)$ such that, for all $x$, $s$, $x$ is constructible from $s$ iff $\langle V, \in\rangle \models \psi(s, x)$. By (a), we can assume $\psi$ is one of the form $\exists v_2\varphi(v_0, v_1, v_2)$ where $\varphi$ is $\Sigma_0$.

If $F$ is primitive recursive with parameter $s \subseteq \omega$, and if $F^{*}L(\alpha) \supseteq Y$, then each element of $Y$ is constructible from $s$. For let $F(x) = G(s,x)$ for all $x$, where $G$ is primitive recursive. It is known that $L(\alpha) \subseteq L_t$ and $s \in D'_t$. By the stability theorem [9; 2.9] for the heirarchy $L_t$, $G(s,x) \in L_t$ for every $x \in L(\alpha)$, i.e., $L_t \supseteq F^{*}L(\alpha) \supseteq Y$.

(5) For use in §2, we define the Prim-closure of a set $Y$ to be the least (in the sense of set inclusion) set $Z \supseteq Y$ such that $Z$ is Prim-closed. By the substitution schema of the definition of primitive recursive functions,

$$Z = \{z \mid z = F(\vec{x}) \text{ for some } \vec{x} \in Y \text{ and some primitive recursive function } F\}.$$

We will only use this terminology when the Prim-closure turns out to be a transitive set.

0.10. The following indicates some of our set theory notation.

$\vec{x}$ represents possible parameters

$0$ empty set

$V$ class of all sets

$\langle x \rangle = x$

$\langle x,y \rangle = \{\{x\},\{x,y\}\}$ ordered pair

$[z]_1$ is $x$ if $z$ is $\langle x,y \rangle$; similarly, $[\langle x,y \rangle]_2 = y$

$\langle x_1,\ldots,x_n \rangle = \langle\langle x_1,\ldots,x_{n-1}\rangle,x_n\rangle$ n-tuple

$Y^{(n)}$ the set of n-tuples with coordinates in $Y$

$\mathrm{dom}(x) = \{z \mid \exists y(\langle z,y \rangle \in x)\}$ domain rng(x) range

$x^{-1}$ inverse of function

$x \upharpoonright y$ the restriction of x to y

$x``y$ the range of $x \upharpoonright y$

fcn(x) x is a function

$x \to y$ denotes function with domain x mapping into y; if it is specifically called partial, then its domain is only required to be a subset of x

$x \mapsto y$ indicates a function f such that f(x) = y

$x \simeq y$ x is cardinally equivalent to y

$\preceq$, $\prec$ the cardinal order relations

trans(x) x is transitive

TC(x) transitive closure (the least transitive $y \supseteq x$)

$\mathrm{Ord}(x) \leftrightarrow \mathrm{trans}(x) \;\&\; \forall yz \in x(y \in z \vee y = z \vee z \in y)$ x is an ordinal

$\alpha$, $\beta$, $\gamma$, $\delta$ denote ordinals

$\alpha+\beta$ ordinal addition

$\mathrm{rank}(x) = \bigcup \{\mathrm{rank}(y)+1 \mid y \in x\}$

$o(X) = \{\alpha \mid \text{ordinal } \alpha \in X\}$

$\omega$ the set of natural numbers 0,1,2,... (these are ordinals)

$\aleph_\alpha$ the $\alpha$-th initial ordinal

$\aleph^+$ the first cardinal $\kappa > \aleph$

$V(\alpha)$ the set of sets of rank less than $\alpha$

$L(\alpha)$ the $\alpha$-th collection of constructible sets (as defined in [9])

$H_{\aleph} = \{x \mid TC(x) \prec \aleph\}$ the set of sets hereditarily of cardinality less than $\aleph$

$\mathcal{P}(x) = \{y \mid y \subseteq x\}$ the power set of $x$

$xy$ depends on context; e.g., if $y$ is considered as a function into the domain of function $x$, then $xy$ is the composition,

$$\langle u,v\rangle \in xy \leftrightarrow \exists w(\langle u,w\rangle \in y \text{ and } \langle w,v\rangle \in x)$$

# SECTION 1

## A FAILURE OF WFT

1.1. MAIN THEOREM. *For every countable admissible ordinal* $\alpha > \omega$, *there exist admissible sets* X *for which WFT fails and* $o(X) = \alpha$.

Let M be a countable admissible set (e.g., $L(\alpha)$) containing $\omega$. It will be shown that an uncountable admissible X exists such that each X-a.i.d. set is countable, $X \supseteq M$, and $o(X) = o(M)$. It follows that WFT fails for X; for consider the X-$\Delta_1$ set $\Gamma$ consisting of the following sentences (1), (2), (3).

1) $\underline{a} \neq \underline{b}$ for each $a \neq b$ of X

2) $\forall uvw(\underline{R}(u,v)\ \&\ \underline{R}(u,w) \rightarrow v \equiv w)$

3) $\bigvee_{\alpha \in \omega} \underline{R}(\underline{\alpha},\underline{a})$ for each a of X

(Compare the $\Gamma$ of the example $H^I_{\aleph_1}$ in 0.7.)

The construction of X uses the following lemma.

1.2. LEMMA. *There exists an indexed set* $\{Mz \mid \text{finite } z \subseteq \aleph_1\}$ *of countable admissible sets for which:*

a) $z \subseteq z'$ *implies that* Mz *is an elementary submodel of* Mz'

b) $Mz \cap Mz' = Mz \cap z'$

c) *there is an indexed set* $\{a_\alpha \mid \alpha \in \aleph_1\}$ *such that:*

$$a_\alpha \in M\{\alpha\};\quad a_\alpha \neq a_\beta \ \textit{for}\ \alpha \neq \beta$$

d) $Mz \supseteq M$ *and* $o(Mz) = o(M)$.

This lemma will eventually be proved in §5. (If M is a model of ZFC, then the proof of Lemma 1.2 will make our M0 a model of the replacement axioms and the axiom of choice. But M0 will not be a model of the power set axiom.)

Continuation of proof of Theorem 1.1: For each $A \subseteq \aleph_1$, define

$$MA =_{\text{df}} \bigcup \{Mz \mid \text{finite } z \subseteq A\}.$$

Then $MA$ is an elementary extension of $M0$ by (a) and a fact of model theory (the union of a directed system of elementary extensions is an elementary extension of each of the members of the system). Therefore, $MA$ is admissible. An uncountable $A$ will be found such that $MA$-a.i.d. sets are countable. By (c) above, $MA$ will be uncountable.

Note that (b) extends itself: if $u, v \subseteq \aleph_1$, then $Mu \cap Mv = Mu \cap v$.

Let $\mathcal{F}$ be the set of uncountable closed subsets $A$ of $\aleph_1$ (closed in the sense of the order topology). Then $[\alpha,\beta]$, $(\alpha,\beta)$, and $[\alpha,\beta)$ denote the closed, open, and half-closed intervals, respectively.

Form a sequence $\langle \varphi_\alpha \mid \alpha \in \aleph_1 \rangle$ of all finitary $M\aleph_1$-sentences $\varphi$ of the definition of a.i.d.

We now define, by induction on $\alpha \in \aleph_1$, a sequence $\langle A_\alpha \mid \alpha \in \aleph_1 \rangle$ where each $A_\alpha \in \mathcal{F}$.

Put $A^o_\alpha = \bigcap \{A_\beta \mid \beta < \alpha\}$ (convention: $\bigcap 0 = \aleph_1$). Then $A^o_\alpha \in \mathcal{F}$, since $\alpha$ is countable (see 0.9 (1)). Take $\gamma_\alpha < \aleph_1$ so that $A^o_\alpha \cap [0,\gamma_\alpha]$ is of order type $\alpha+1$. There are two cases for the definition of $A_\alpha$.

<u>Case 1</u>. Suppose there exists $A$ such that $A \in \mathcal{F}$, $A \cap [0,\gamma_\alpha] = A^o_\alpha \cap [0,\gamma_\alpha]$, and $\varphi_\alpha$ is a $MA$-a.i. definition of some uncountable set $R$. Since $R$ is uncountable and $M[0,\gamma_\alpha]$ is countable, there are $y$, $z$, $\gamma$ such that $y \in R - M[0,\gamma_\alpha]$, finite $z \subseteq A$, $y \in Mz$, and $\gamma = \overline{\sup}\, z$ (i.e., $\gamma = \bigcup \{\alpha+1 \mid \alpha \in z\}$ is the strict least upper bound). Define

$A_\alpha = A-(\gamma_\alpha,\gamma) \in \mathcal{F}$ for a choice of some such $A$, $R$, $y$, $z$, $\gamma$.

Case 2. Suppose no $A$ exists for case 1. Define $A_\alpha = A^o_\alpha \in \mathcal{F}$. Put

$$A_{\aleph_1} = \cap \{A_\alpha \mid \alpha < \aleph_1\}.$$

Note that, if $\alpha < \aleph_1$, then $A_{\aleph_1} \cap [0,\gamma_\alpha] = A^o_\alpha \cap [0,\gamma_\alpha]$. To prove this, put $A^o_{\aleph_1} = A_{\aleph_1}$. We prove by induction on $\beta \in (\alpha,\aleph_1]$ that $A^o_\beta \cap [0,\gamma_\alpha] = A^o_\alpha \cap [0,\gamma_\alpha]$. Let $\beta \in (\alpha,\aleph_1]$. Then

$$(*) \qquad A^o_\beta \cap [0,\gamma_\alpha] = A^o_\alpha \cap \bigcap_{\delta\in[\alpha,\beta)} A_\delta \cap [0,\gamma_\alpha].$$

For any $\delta \in [\alpha,\beta)$, $A^o_\delta \cap [0,\gamma_\alpha] = A^o_\alpha \cap [0,\gamma_\alpha]$, either by $\delta = \alpha$ or by induction hypothesis. These equal sets have the same order type $\alpha+1$. At induction step $\delta \geq \alpha$ of the construction, we chose $\gamma_\delta$ such that $A^o_\delta \cap [0,\gamma_\delta]$ is of order type $\delta+1 \geq \alpha+1$. Therefore, $A^o_\delta \cap [0,\gamma_\delta] \supseteq A^o_\delta \cap [0,\gamma_\alpha]$ and $[0,\gamma_\delta] \supseteq [0,\gamma_\alpha]$. The definition of $A_\delta$ guaranteed that $A_\delta \cap [0,\gamma_\delta] = A^o_\delta \cap [0,\gamma_\delta]$. Combining, $A_\delta \cap [0,\gamma_\delta] = A_\delta \cap [0,\gamma_\delta] \cap [0,\gamma_\alpha] = A^o_\delta \cap [0,\gamma_\alpha] = A^o_\alpha \cap [0,\gamma_\alpha]$. By this equality and (*), the induction is finished.

Consider $A_{\aleph_1}$. It is closed. It is uncountable: for each $\alpha < \aleph_1$, it has the segment $A_{\aleph_1} \cap [0,\gamma_\alpha] = A^o_\alpha \cap [0,\gamma_\alpha]$ of order type $\alpha+1$. Thus, $A_{\aleph_1} \in \mathcal{F}$. Put $X = MA_{\aleph_1}$.

We show that no $\varphi_\alpha$ is a X-a.i. definition of an uncountable set. Suppose that some $\varphi_\alpha$ is a X-a.i. definition of some uncountable set R. Since $A_{\aleph_1} \in \mathcal{F}$ and $A_{\aleph_1} \cap [0,\gamma_\alpha] = A^o_\alpha \cap [0,\gamma_\alpha]$, $A_{\aleph_1}$ satisfies the supposition of case 1 at induction step $\alpha$. At induction step $\alpha$, we chose $A$, $R'$, $y$, $z$, $\gamma$ such that $\varphi_\alpha$ is a MA-a.i. definition of $R'$, $y \in R'$, $y \notin M[0,\gamma_\alpha]$, $y \in Mz$, and $\gamma = \overline{\sup}\ z$. Then $y \notin M(\aleph_1-(\gamma_\alpha,\gamma))$: otherwise, since

$$(\aleph_1-(\gamma_\alpha,\gamma)) \cap z \subseteq (\aleph_1-(\gamma_\alpha,\gamma)) \cap [0,\gamma) = [0,\gamma_\alpha],$$

$y \in M(\aleph_1-(\gamma_\alpha,\gamma)) \cap Mz = M((\aleph_1-(\gamma_\alpha,\gamma)) \cap z) \subseteq M[0,\gamma_\alpha]$, by the extension of (b); this contradicts $y \notin M[0,\gamma_\alpha]$. Thus, $y \notin M(\aleph_1-(\gamma_\alpha,\gamma)) \supseteq MA_\alpha \supseteq X \supseteq R$. Thus, $R \neq R'$, and $MA \supseteq MA_\alpha \supseteq X$. This contradicts the persistence property.

Since every possible X-a.i. definition is some $\varphi_\alpha$, we are done.

## SECTION 2

## EXAMPLE WITH INDIVIDUALS

Theorem 1.1 has a simpler proof when individuals are allowed to occur both in set theory and in admissible sets. We describe this proof since some of this proof has an analog in our proof of Lemma 1.2. Acutally, we will now prove (with individuals) a stronger version of Theorem 1.1 that corresponds to the consistency of 6.8 mentioned in the introduction. The notation introduced in this section is only for this section; it conflicts with later usage. The rest of this paper does not logically depend on this section.

Let $\aleph$ be an uncountable cardinal. let $I = \{i_{\alpha,n} \mid \alpha\in\aleph,\ n\in\omega\}$ be an $\aleph\times\omega$-indexed set of distinct individuals. For each $\alpha\in\aleph$, let $a_\alpha = \{\langle n, i_{\alpha,n}\rangle \mid n\in\omega\}$. Let $M$ be an admissible set containing $\omega$ and not containing individuals. For each $u\subseteq\aleph$, let $Tu$ be the Prim-closure of $M\cup\{a_\alpha \mid \alpha\in u\}$ defined in 0.9 (5). (An $Mz$ of Lemma 1.2 is analogous to $T\omega\cup z$.) We will show that $Tu$ is admissible. Afterwards, we will show that, for uncountable $A\subseteq\aleph$ such that $\aleph - A$ is infinite, WFT fails for $TA$.

(If $M$ is some segment $L(\alpha)$ of the constructible hierarchy and if $z$ is a finite subset of $\aleph$, then it turns out that

$$Tz = L_{TC(\{a_\beta \mid \beta\in z\})}(\alpha),$$

where $L_t$ was mentioned in 0.9 (4). Also, $Tu = \bigcup\{Tz \mid \text{finite } z\subseteq u\}$.)

2.1. LEMMA. Symmetry.

1) *A* 1-1 *onto* $g\colon u\to v$ *determines an* $\in$*-isomorphism* $g\colon Tu\to Tv$ *extending* $i_{\alpha,n}\mapsto i_{g(\alpha),n}$.

*This also determines an extended map* $g$ *of the relations of* $Tu$ *to those of* $Tv$. *For this map,*

$$\langle Tu,\in,R,S\rangle \models \varphi(\vec{x}) \quad \textit{iff} \quad \langle Tv,\in,gR,gS\rangle \models \varphi(g\vec{x}).$$

*For* $z \subseteq u$, $(g \restriction z): Tz \to Tg''z$ *is* $g \restriction Tz$. *If* $g$ *is identity, then* $g: Tu \to Tv$ *is identity.*

2) *Consider* $\gamma \in \aleph$, *nonempty* 1-1 *finite sequence* $s: m \to \aleph$, *and* $z \subseteq \aleph$ *such that* $\gamma \notin z$ *and* $z \cap \mathrm{rng}(s) = 0$. *These determine an* $\in$-*isomorphism* $\underline{\gamma s z}: T(z \cup \mathrm{rng}(s)) \to T(z \cup \{\gamma\})$ *extending*

$$i_{s(r),n} \mapsto i_{\gamma,mn+r} \qquad \textit{for} \quad r \in m, \quad n \in \omega$$

$$i_{\alpha,n} \mapsto i_{\alpha,n} \qquad \textit{for} \quad \alpha \in z, \quad n \in \omega.$$

PROOF: A 1-1 onto map $h$ of individuals determines an $\in$-isomorphism $\underline{h}$ of all the sets built up from the corresponding individuals, by induction on $\in$:

$\underline{h}(i) = h(i)$ for $i \in \mathrm{dom}(h)$

$\underline{h}(x) = \underline{h}''x$ for $x \notin \mathrm{dom}(h)$ but all individuals of $TC(x)$ are in $\mathrm{dom}(h)$.

It can be shown that $\underline{h}(F(\vec{x})) = F(\underline{h}(\vec{x}))$ for every primitive recursive function $F$ (by induction on the defining schemes for primitive recursive functions; or by knowing that $\underline{h}: U \to W$ is an $\in$-isomorphism of inner models of ZF with individuals, i.e., of Prim-closed classes, and by applying the absoluteness property [9; 2.3 (4)] of the defining formula $\varphi$ of $F$:

$$\langle V,\in\rangle \models \varphi(\vec{x},F(\vec{x})) \quad \text{implies} \quad \langle U,\in\rangle \models \varphi(\vec{x},F(\vec{x})) \quad \text{implies}$$
$$\langle W,\in\rangle \models \varphi(\underline{h}\vec{x},\underline{h}F(\vec{x})) \quad \text{implies} \quad \langle V,\in\rangle \models \varphi(\underline{h}\vec{x},\underline{h}F(\vec{x})) \quad \text{implies}$$
$$\underline{h}F(\vec{x}) = F(\underline{h}\vec{x}).)$$

For (1), we show $g''Tu \subseteq Tv$. For every $y \in Tu$, there are $\alpha_1,\ldots,\alpha_n \in u$, $x \in M$, and primitive recursive function $F$ such that $y = F(a_{\alpha_1},\ldots,a_{\alpha_n},x)$. Then

$gF(a_{\alpha_1},\ldots,a_{\alpha_n},x) = F(ga_{\alpha_1},\ldots,ga_{\alpha_n},gx) = F(a_{g(\alpha_1)},\ldots,a_{g(\alpha_n)},x) \in Tv.$

By considering $g^{-1}$, we also have $g``Tu \supseteq Tv$.

For (2), we show $\underline{\gamma sz}``T(z \cup rng(s)) \subseteq T(z \cup \{\gamma\})$. Abbreviate by putting $\tau = \underline{\gamma sz}$. For each $r \in m$, define primitive recursive $G_r$ so that

$$G_r(a_\gamma) = \{\langle n, a_\gamma(mn+r)\rangle \mid n \in \omega\} = \{\langle n, i_{\gamma,mn+r}\rangle \mid n \in \omega\}$$

$$= \{\langle n, \tau(i_{s(r),n})\rangle \mid n \in \omega\} = \tau a_{s(r)}.$$

Consider any $\alpha_1,\ldots,\alpha_n \in z$ and any $x \in M$. Consider any primitive recursive function $F$ with $n+m+1$ arguments. Then

$$\tau F(a_{\alpha_1},\ldots,a_{\alpha_n},a_{s(0)},\ldots,a_{s(m-1)},x)$$

$$= F(\tau a_{\alpha_1},\ldots,\tau a_{\alpha_n},\tau a_{s(0)},\ldots,\tau a_{s(m-1)},\tau x)$$

$$= F(a_{\alpha_1},\ldots,a_{\alpha_n},G_0(a_\gamma),\ldots,G_{m-1}(a_\gamma),x) \in T(z \cup \{\gamma\}).$$

For (2), we show $\tau^{-1}``T(z \cup \{\gamma\}) \subseteq T(z \cup rng(s))$. Define primitive recursive function $G_r$ so that $G_r(a_{s(r)}) = \{\langle mn+r, a_{s(r)}(n)\rangle \mid n \in \omega\}$. Then for $\alpha_1,\ldots,\alpha_n \in z$, $x \in M$, and primitive recursive $F$:

$$\tau^{-1}F(a_{\alpha_1},\ldots,a_{\alpha_n},a_\gamma,x)$$

$$= F(\tau^{-1}a_{\alpha_1},\ldots,\tau^{-1}a_{\alpha_n},\tau^{-1}a_\gamma,\tau^{-1}x)$$

$$= F(a_{\alpha_1},\ldots,a_{\alpha_n}, \cup\{G_0(a_{s(0)}),\ldots,G_{m-1}(a_{s(m-1)})\},x) \in T(z \cup rng(s)).$$

2.2 LEMMA. *For every finite* $z \subseteq \aleph$, *transitive* $Tz$ *satisfies the* $\Sigma_0$*-reflection principle. Also for finite* $z' \subseteq \aleph$, $Tz \cap Tz' = Tz \cap z'$.

PROOF: By the last lemma, we can assume

$$z = \{\alpha_1,\ldots,\alpha_n\} \subseteq \omega.$$

Define inductively the inner model $\langle Nz, E\rangle$ of $M$ so that

$$x \in Nz \longleftrightarrow x \in z \times \omega \times \{2\} \quad \text{or} \quad x \subseteq Nz \times \{1\} \qquad (\text{and } x \in M)$$

$$x\,E\,y \longleftrightarrow \langle x,1\rangle \in y \qquad (\text{and } x,y \in Nz).$$

It will be shown that $Nz$ is a model of $\Sigma_0$-reflection and is ε-isomorphic to $Tz$.

The relations $Nz$ and $E$ are restrictions to $M$ of primitive recursive relations; thus, $Nz$ and $E$ are $M$-$\Delta_1$. For the rest of the proof of this lemma, $\langle Nz,E\rangle \models \varphi(\vec{x})$ is written $\models \varphi(\vec{x})$. Also, $\varphi$ will represent an arbitrary $\Sigma_0$ formula. Then $\models \varphi(\vec{x})$ is a $M$-$\Sigma_0$ relation of the arguments $\vec{x}$, since the quantifier "$\forall v \in y$" translates to "$\forall x(\langle x,1\rangle \in y \rightarrow\ )$."

The proof that $Nz$ is a model of the axioms of admissibility is now indicated. Here $x, \vec{x}, y$ are assumed in $Nz$.

Pairs: $\{\langle x,1\rangle,\langle y,1\rangle\} \in Nz$.

Unions: $\{\langle y,1\rangle \mid \exists z(\langle z,1\rangle \in x \;\&\; \langle y,1\rangle \in z)\} \in Nz$.

$\Sigma_0$-separation: $\{\langle y,1\rangle \mid \langle y,1\rangle \in x \;\&\; \models \varphi(y,\vec{x})\} \in Nz$.

$\Sigma_0$-reflection: Assume $\models \forall v_0 \in x\, \exists v_1 \varphi(\vec{x},v_0,v_1)$; we show that there is $x_2 \in Nz$ such that $\models \forall v_0 \in x\, \exists v_1 \in x_2\, \varphi(\vec{x},v_0,v_1)$. We can assume that $x \subseteq Nz \times \{1\}$. Then $\forall x_0 \in \mathrm{dom}(x) \exists x_1 \in Nz(\models \varphi(\vec{x},x_0,x_1))$. By $\Sigma_1$-reflection for $M$, since $Nz$ is $M$-$\Delta_1$, there is a restriction $w$ in $M$ for "$\exists x_1$"; i.e., $\forall x_0 \in \mathrm{dom}(x)\exists x_1 \in w(\models \varphi(\vec{x},x_0,x_1))$. Put $x_2 = (w \cap Nz) \times \{1\} \in Nz$. This is the required $x_2$.

Define inductively $\check{x} = \{\langle \check{y},1\rangle \mid y \in x\}$. Then $\check{x} \in Nz$ for any $x \in M$. Define $\check{\{x,y\}}$ to be the $\{\langle x,1\rangle,\langle y,1\rangle\}$ above. Define $\check{\langle x,y\rangle}$ to be $\check{\{}\check{\{}x\check{\}},\check{\{}x,y\check{\}}\check{\}}$, the "ordered pair" of $Nz$. For $\alpha \in z$, define $b_\alpha$ to be $\{\check{\langle}\check{m},\langle\alpha,m,2\rangle\check{\rangle} \mid m \in \omega\}$. Since $\omega \in M$, $b_\alpha \in Nz$.

Define the primitive recursive function $F$ so that

$$F(a_{\alpha_1},\ldots,a_{\alpha_n},x) = \begin{cases} a_{\alpha_1}(m) & \text{if } x = \langle\alpha_1,m,2\rangle \text{ and } m \in \omega \\ \vdots & \vdots \\ a_{\alpha_n}(m) & \text{if } x = \langle\alpha_n,m,2\rangle \text{ and } m \in \omega \\ \{F(a_{\alpha_1},\ldots,a_{\alpha_n},y) \mid \langle y,1\rangle \in x\} & \text{otherwise.} \end{cases}$$

Considering $a_{\alpha_1},\ldots a_{\alpha_n}$ as parameters, $F$ is an $\varepsilon$-isomorphism of $Nz$ onto $F``Nz$. For $x \in M$, $F(\check{x}) = x$. For $\alpha \in z$, $F(b_\alpha) = a_\alpha$. Thus $F``Nz \supseteq M \cup \{a_{\alpha_1},\ldots,a_{\alpha_n}\}$. Now $F``Nz$ is transitive and (since $F$ is an isomorphism) satisfies the axioms of admissibility; thus $F``Nz$ is Prim-closed. Thus, $Tz \subseteq F``Nz$. Since $M \subseteq Tz$ and $Tz$ is closed under $F$, $F``Nz \subseteq F``M \subseteq Tz$. Thus, $F``Nz = Tz$.

$Tz \cap Tz' = Tz \cap z'$: For $z_1 \subseteq z$, let $F_1: Nz_1 \longrightarrow Tz_1$ be the corresponding $\varepsilon$-isomorphism. By induction, for $x \in Nz$: $F(x)$ is built up from elements of $\{i_{\alpha,n} \mid \alpha \in z_1,\ n \in \omega\}$ iff $x \in Nz_1$ and $F(x) = F_1(x) \in Tz_1$. Put $z_1 = z \cap z'$.

2.3 LEMMA. *The set* $Tu$ *is admissible and* $o(Tu) = o(M)$.

PROOF: $o(Tu) = o(M)$: This is by observing the $F$ of the last lemma; or this follows from the general result of [9; 2.6].

Since $Tu$ is Prim-closed, we only need to show $\Sigma_0$-reflection (and transitivity). By the last lemma, we can assume $u$ is not finite. Note that $Tu = \bigcup \{Tz \mid \text{finite } z \subseteq u\}$ since a primitive recursive function has only finitely many arguments. (Thus, $Tu$ is transitive.) We show by symmetry that the $\Sigma_0$-reflection principle for $Tu$ follows from the $\Sigma_0$-reflection for each $Tz$ such that $z$ is finite.

Suppose $x,\vec{x} \in Tu$ and $\varphi$ is a $\Sigma_0$ formula. By the absoluteness property of $\Sigma_0$ formulas, we can write $\models \varphi(\vec{y})$ for any $\langle Tv,\epsilon\rangle \models \varphi(\vec{y})$, provided $\vec{y} \in Tv$. Suppose $\forall x_0 \in x \exists x_1 \in Tu(\models \varphi(\vec{x},x_0,x_1))$. Find finite $z \subseteq u$ such that $x,\vec{x} \in Tz$. Find $\gamma \in u-z$. We will show that

$$\forall x_0 \in x \exists x_1 \in T(z \cup \{\gamma\})(\models \varphi(\vec{x},x_0,x_1)).$$

By $\Sigma_0$-reflection for $T(z \cup \{\gamma\})$, it will then follow that there exists $w \in T(z \cup \{\gamma\}) \subseteq Tu$ such that

$$\forall x_0 \in x \exists x_1 \in w(\models \varphi(\vec{x},x_0,x_1)).$$

Assume $x_0 \in x$. For some $x_1 \in Tu$, $\vdash \varphi(\vec{x},x_0,x_1)$. Find finite $z' \subseteq u$ such that $x_1 \in Tz'$. We can assume $z' - z \neq 0$. List $z' - z$ by a nonempty 1-1 finite sequence $s$. By symmetry (2), take $\tau = \gamma sz$. Then $\models \varphi(\tau\vec{x},\tau x_0,\tau x_1)$, since $\varphi$ is $\Sigma_0$ and $\tau$ is an $\in$-isomorphism. Then $\tau\vec{x} = \vec{x}$; $\tau x_0 = x_0$; $\models \varphi(\vec{x},x_0,\tau x_1)$; $\tau x_1 \in T(z \cup \{\gamma\})$. Thus, $\forall x_0 \in x \exists x_1 \in T(z \cup \{\gamma\})(\models \varphi(\vec{x},x_0,x_1))$. QED

Incidental Remark. For infinite $u \subseteq v \subseteq \aleph$, $Tu$ is an elementary submodel of $Tv$. By the fact of model theory used in the proof of Theorem 1.1, we can assume $u \simeq \omega \simeq v$. Given $\vec{x} \in Tu$ and formula $\varphi(\vec{x})$, find finite $z$ such that $\vec{x} \in Tz$. Then find 1-1 onto $g: u \to v$ such that $g \upharpoonright z$ is identity. Then

$$\langle Tu,\in\rangle \models \varphi(\vec{x}) \quad \text{iff} \quad \langle gTu,\in\rangle \models \varphi(\vec{x}) \quad \text{iff} \quad \langle Tv,\in\rangle \models \varphi(\vec{x}).$$

Let $A$ be an uncountable subset of $\aleph$ such that $\aleph - A$ is infinite. (Actually, the next theorem holds for any uncountable subset $A$ of $\aleph$.)

2.4. THEOREM. *The set* $TA$ *is an admissible set which fails WFT. Also,* $M \subseteq TA$ *and* $o(M) = o(TA)$. *For any admissible ordinal* $\alpha > \omega$, *we could have put* $M = L(\alpha)$, *so that* $o(TA) = \alpha$.

PROOF: The main point is the following lemma.

2.5. LEMMA. *Every* $TA$-*a.i.d. set is a subset of some* $Tz$ *where* $z$ *is finite.*

PROOF: Let $\varphi(\vec{x})$ be a $TA$-a.i. definition of $R$. Then $\vec{x}$ is contained in some Tz where finite $z \subseteq A$. For some S,

$$\langle TA, \in, R, S\rangle \models \varphi(\vec{x}).$$

Suppose $R \not\subseteq Tz$. There is $y \in R - Tz$. For some finite $z' \supseteq z$, $y \in Tz'$. Find 1-1 onto $g: A \to B$ such that $A \subseteq B \subseteq \aleph$, $g \restriction z$ is identity, and $g''(z'-z) \subseteq \aleph - A$. Take the g of symmetry (1). Then

$$\langle TB, \in, gR, gS\rangle \models \varphi(g\vec{x});$$

transitive $TB \supseteq TA$; and $g\vec{x} = \vec{x}$. By persistence, $gR = R$. But $gy \in gR \cap gTz' \subseteq R \cap gTz' \subseteq TA \cap Tg''z' \subseteq T(A \cap g''z') \subseteq Tz$. This contradicts $gy \notin gTz = Tz$. QED Lemma

For the theorem, let $\Gamma$ be as in the example $H^{I}_{\aleph_1}$ of 0.7, replacing $H^{I}_{\aleph_1}$ by TA.

# SECTION 3

## INNER BOOLEAN MODELS FOR ADMISSIBLE SETS

We assume familiarity with complete Boolean algebras and Boolean algebraic models for set theory (see, e.g., [25], [22], and [23]). We here consider a complete Boolean algebra to be a special kind of partial ordering $\langle\mathbb{B},\leq\rangle$. The other operations are definable from $\leq$. The complement of $b$ is written $\neg b$. For $X \subseteq \mathbb{B}$, the greatest lower bound $\bigwedge X$ (or meet) of $X$ exists, by definition of "complete." The least upper bound $\bigvee X$ (or join) is equal to $\neg\bigwedge\{\neg b \mid b\in X\}$. For $b,c \in \mathbb{B}$, $b\wedge c = \bigwedge\{b,c\}$ and $b\vee c = \bigvee\{b,c\}$. The zero is $\mathbb{0}$ and the unit is $\mathbb{1}$. We assume $\mathbb{0} \neq \mathbb{1}$. We write $\mathbb{2}$ for the subalgebra $\{\mathbb{0},\mathbb{1}\}$. We notationally identify $\mathbb{B}$ with $\langle\mathbb{B},\leq\rangle$. In this section $p$, $q$, $r$, $s$ denote members of any given partially ordered set $\mathbb{P}$.

A complete subalgebra $\mathbb{B}_1$ of $\mathbb{B}$ is a substructure of $\mathbb{B}$ such that

$$\forall b \in \mathbb{B}_1((\neg b) \in \mathbb{B}_1)$$

and

$$\forall X \subseteq \mathbb{B}_1(\bigwedge X \in \mathbb{B}_1).$$

Then $\mathbb{B}_1$ is a complete Boolean algebra with complement and meet given by those of $\mathbb{B}$.

3.1. A countable admissible set does not contain infinite complete Boolean algebras [25]. However, it does contain dense subsets of some infinite complete Boolean algebras (when it contains $\omega$). In order to make this fact useful, we now list some known elementary relationships between dense sets, complete Boolean algebras, and isomorphisms. There is a connection with what is commonly called weak forcing.

A subset $\mathbb{P}$ of $\mathbb{B}$ is called dense if

$$\mathbb{0} \notin \mathbb{P}$$

and

$$\forall b \in \mathbb{B} \exists p \in \mathbb{P}(b \neq \mathbb{0} \longrightarrow p \leq b).$$

(Trivial example: $\mathbb{B} - \{\mathbb{0}\}$ is dense in $\mathbb{B}$.)

A partial ordering $\langle \mathbb{P}, \leq \rangle$ is a dense substructure of some Boolean algebra iff $\forall pq(p \nleq q \longrightarrow \exists r \leq p \; \forall s \leq r(s \nleq q))$. (For this fact, [25; 12B] refers to [5]. If $\mathbb{P}$ is dense in some $\mathbb{B}$, and if $p \nleq q$, then $p \wedge \neg q \neq \mathbb{0}$, so there is some $r \leq p \wedge \neg q$; the converse is proven by considering the $\mathbb{G}$ below.) The (minimal) completion of the Boolean algebra is unique up to an isomorphism, and $\mathbb{P}$ is dense in the completion [25; 35.2].

For the rest of this section, $\mathbb{P}$ denotes a dense subset of a complete $\mathbb{B}$. Note the contrapositive of an extension of the condition for being a dense set:

$$\forall p \forall b \in \mathbb{B}(\forall r \leq p \exists s \leq r(s \leq b) \longrightarrow p \leq b).$$

The following is a slight modification of what is in [5] or [8; pp. 100-103].

Define:

$$\neg S = \{p \mid \forall q \leq p(q \notin S)\} \qquad \text{for } S \subseteq \mathbb{P};$$

$$\mathbb{G} = \{S \mid S \subseteq \mathbb{P} \text{ and } S = \neg\neg S\}.$$

Then $\mathbb{G}$ is ordered by inclusion $\subseteq$. For this ordering, $\mathbb{G}$ is a complete Boolean algebra (this is provable from the fact that $\langle \mathbb{P}, \leq \rangle$ is a partial ordering). Further,

$$p \longmapsto \{q \mid q \leq p\} \qquad (= \neg\neg\{q \mid q \leq p\} \in \mathbb{G}, \text{ by the condition for being a dense set})$$

is a $\leq$-embedding $\mu\colon \mathbb{P} \longrightarrow \mathbb{G}$ onto a dense subset of $\mathbb{G}$. The complement of an element $S$ of $\mathbb{G}$ is the $\neg S$ just defined. For each $X \subseteq \mathbb{G}$, the meet $\wedge X$ is $\cap X$, and the join $\vee X$ is $\neg\neg\cup X$.

Then $\mathbb{B}$ is isomorphic with $\mathcal{B}$. The unique isomorphism $\mu: \mathbb{B} \to \mathcal{B}$ extending $\mu: \mathbb{P} \to \mathcal{B}$ is given by

$$\mu(b) = \{p \mid p \le b\};$$

and

$$\mu^{-1}(S) = \vee S \quad \text{for } S \in \mathcal{B}.$$

The following uniquely determine the operations $\neg, \wedge, \vee$ of $\mathbb{B}$:

$$\forall p(p \le \neg b \leftrightarrow \forall q \le p(q \not\le b))$$

$$\forall p(p \le \wedge X \leftrightarrow \forall b \in X(p \le b))$$

$$\forall p(p \le \vee X \leftrightarrow \forall q \le p \exists r \le q \exists b \in X(r \le b)).$$

This can be proved directly or by using $\mu$ above. For example,

$$\neg b = \vee\{p \mid p \le \neg b\}$$

and

$$p \le \neg b \leftrightarrow p \in \mu(\neg b) \leftrightarrow p \in \neg\mu(b) \leftrightarrow \forall q \le p(q \notin \mu(b)) \leftrightarrow \forall q \le p(q \not\le b).$$

Let $\mathbb{P}_1$ and $\mathbb{P}_2$ be dense subsets of complete Boolean algebras $\mathbb{B}_1$ and $\mathbb{B}_2$, respectively; let $\tau: \mathbb{P}_1 \to \mathbb{P}_2$ be a $\le$-isomorphism. Then there is a unique isomorphism $\tau: \mathbb{B}_1 \to \mathbb{B}_2$ that extends $\tau$ [25; 12.5, 33.4]. Consider the embeddings $\mu_1: \mathbb{P}_1 \to \mathcal{B}_1$, $\mu_2: \mathbb{P}_2 \to \mathcal{B}_2$ and their extensions $\mu_1: \mathbb{B}_1 \to \mathcal{B}_1$, $\mu_2: \mathbb{B}_2 \to \mathcal{B}_2$ (where $\mu_1(b) = \{p \in \mathbb{P}_1 \mid p \le b\}$ for $b \in \mathbb{B}_1$, etc.). Then $\mu_2\tau\mu_1^{-1}: \mathcal{B}_1 \to \mathcal{B}_2$ is the unique isomorphism which extends $\mu_2\tau\mu_1^{-1}: \mu_1``\mathbb{P}_1 \to \mu_2``\mathbb{P}_2$. Note that for $p \in \mathbb{P}_2$, for $S \in \mathcal{B}_1$,

$$p \in \mu_2\tau\mu_1^{-1}(S) \leftrightarrow p \le \tau\mu_1^{-1}(S) \leftrightarrow \tau^{-1}p \le \mu_1^{-1}(S) \leftrightarrow \tau^{-1}p \in S \leftrightarrow p \in \tau``S.$$

Thus, $\mu_2\tau\mu_1^{-1}(S) = \tau``S$. Then $\mu_2\tau\mu_1^{-1}$ is primitive recursive with parameter $\tau: \mathbb{P}_1 \to \mathbb{P}_2$.

Suppose $\mathbb{P}_1 \subseteq \mathbb{P}$ and $\mathbb{P}_1$ is dense in a complete subalgebra $\mathbb{B}_1$ of $\mathbb{B}$. Then $\mu\mu_1^{-1}: \mathcal{B}_1 \longrightarrow \mathcal{B}_1^*$ is the unique extending isomorphism of $\mu\mu_1^{-1}: \mu_1``\mathbb{P}_1 \longrightarrow \mu``\mathbb{P}_1$ onto some complete subalgebra $\mathcal{B}_1^*$ of $\mathcal{B}$. Note that, for $S \in \mathcal{B}_1$,

$$\mu\mu_1^{-1}(S) = \mu(\vee S) = \{p \mid p \leq \vee S\}$$

$$= \{p \mid \forall q \leq p \exists r \leq q \exists s \in S(r \leq s)\}$$

and, for $S \in \mathcal{B}_1^*$,

$$(\mu\mu_1^{-1})^{-1}(S) = \mu_1(\vee S) = \{p \in \mathbb{P}_1 \mid p \leq \vee S\} = \mathbb{P}_1 \cap S.$$

Thus, $\mu\mu_1^{-1}$ and its inverse are both primitive recursive with parameters $<\mathbb{P},\leq>$ and $\mathbb{P}_1$.

We now define, for each isomorphism $\tau$ of dense subsets of complete subalgebras of $\mathbb{B}$, an isomorphism $\tau^*$ of complete subalgebras of $\mathcal{B}$. Suppose that $\mathbb{P}_1,\mathbb{P}_2 \subseteq \mathbb{P}$ and $\mathbb{P}_1$, $\mathbb{P}_2$ are dense subsets of complete subalgebras $\mathbb{B}_1$, $\mathbb{B}_2$ (respectively) of $\mathbb{B}$. Suppose that $\tau: \mathbb{P}_1 \longrightarrow \mathbb{P}_2$ is an isomorphism with extension $\tau: \mathbb{B}_1 \longrightarrow \mathbb{B}_2$, as above. Then we define $\tau^*: \mathcal{B}_1^* \longrightarrow \mathcal{B}_2^*$ by

$$\tau^* = (\mu\mu_2^{-1})(\mu_2\tau\mu_1^{-1})(\mu\mu_1^{-1})^{-1},$$

i.e.,

$$\tau^* = \mu\tau(\mu^{-1} \restriction \mathcal{B}_1^*).$$

By the above, $\tau^*$ is primitive recursive with parameters $\tau: \mathbb{P}_1 \longrightarrow \mathbb{P}_2$ and $<\mathbb{P},\leq>$. If $\mathbb{P}_0 \subseteq \mathbb{P}_1$ and $\mathbb{P}_0$ is also a dense subset of some complete subalgebra of $\mathbb{B}$, then

$$(\tau \restriction \mathbb{P}_0)^* = \tau^* \restriction \mathcal{B}_0^*.$$

If $\sigma: \mathbb{P}_2 \longrightarrow \mathbb{P}_3$ is another isomorphism where $\mathbb{P}_3 \subseteq \mathbb{P}$ and $\mathbb{P}_3$ is dense in some complete subalgebra of $\mathbb{B}$, then

$$(\tau\sigma)^* = \tau^*\sigma^*.$$

If $\tau$ is identity, then $\tau^*$ is dentity on $\mathbb{B}_1^*$. It follows from the last two sentences that

$$(\tau^{-1})^* = (\tau^*)^{-1}.$$

3.2. We now describe our terminology and some basic facts about Boolean structures and models.

A Boolean structure $\mathcal{U} = \langle U, R_0, \ldots, R_m \rangle$ of similarity type $(n_0, \ldots, n_m)$ consists of a complete Boolean algebra $\mathbb{B}$, a nonempty set $U$, and, for each $i \leq m$, a function $R_i : U^{(n_i)} \longrightarrow \mathbb{B}$.

Each nonempty subset $U_1$ of $U$ determines a Boolean substructure of $\mathcal{U}$, namely $\langle U_1, R_0 \restriction U_1^{(n_0)}, \ldots, R_m \restriction U_1^{(n_m)} \rangle$ with the Boolean algebra $\mathbb{B}$ or with some complete subalgebra $\mathbb{B}_1$ of $\mathbb{B}$ such that each $R_i \restriction U_1^{(n_i)}$ maps into $\mathbb{B}_1$.

An isomorphism $\mathcal{U} \longrightarrow \mathcal{U}'$ of Boolean structures of the same similarity type consists of some $\tau$ and $\underline{\tau}$ such that (letting $'$ denote the corresponding entity of $\mathcal{U}'$):

$$\tau : \mathbb{B} \longrightarrow \mathbb{B}' \quad \text{is a} \quad \leq\text{-isomorphism}$$

$$\underline{\tau} : U \longrightarrow U' \quad \text{is an} \quad 1\text{-}1 \quad \text{onto function;}$$

and

$$\tau(R_i(x_1, \ldots, x_{n_i})) = R_i'(\underline{\tau}(x_1), \ldots, \underline{\tau}(x_{n_i})) \qquad \text{for all } i \leq m, \text{ all } x_1, \ldots, x_{n_i} \in U.$$

For the above $\tau$, $\underline{\tau}$, and $U_1 \subseteq U$, there is the Boolean structure isomorphism $\tau \restriction \mathbb{B}_1$, $\underline{\tau} \restriction U_1$ of the substructures $U_1$ and $\underline{\tau}``U_1$.

Consider the (finitary) language with the relation symbols $\underline{R}_0, \ldots, \underline{R}_m$, where $\underline{R}_i$ is $n_i$-ary for $i \leq m$. Then $\mathcal{U}$ can be considered an interpretation of this language.

For each formula $\varphi(\vec{v})$ of the language, and for $\vec{x} \in U$, the Boolean value $[\![\varphi(\vec{x})]\!]$ of the $U$-sentence is defined inductively:

$$[\![\underline{R}_i(x_1,\ldots,x_{n_i})]\!] = R_i(x_1,\ldots,x_{n_i}) \quad \text{for } i \le m$$

$$[\![\neg\varphi]\!] = \neg[\![\varphi]\!]$$

$$[\![\varphi \,\&\, \psi]\!] = [\![\varphi]\!] \wedge [\![\psi]\!]$$

$$[\![\forall v\varphi(v)]\!] = \bigwedge\{[\![\varphi(x)]\!] \mid x \in U\}.$$

An U-sentence $\varphi$ is Boolean valid for the structure $\mathcal{U}$ iff $[\![\varphi]\!] = \mathbb{1}$. Also, $\mathcal{U}$ is a Boolean model of a set $\Phi$ of sentences iff each member of $\Phi$ is Boolean valid. The main point of a Boolean model $\mathcal{U}$ is that: every axiom of predicate logic is Boolean valid; every rule of inference of predicate logic takes Boolean valid premises to a Boolean valid conclusion. If some $\underline{R}_j$ is the equality symbol, then we will require that the equality axioms are Boolean valid.

For an isomorphism $\tau$, $\underline{\tau}$ of Boolean structures, one proves by induction that

$$\tau[\![\varphi(\vec{x})]\!] = [\![\varphi(\underline{\tau}\vec{x})]\!].$$

3.3 We now list some definitions and facts from [22]. One defines recursively the (improper) Boolean structure $\langle V^{(\mathbb{B})}, [\![\varepsilon]\!], [\![\equiv]\!]\rangle$:

$$x \in V^{(\mathbb{B})} \leftrightarrow x \text{ is a partial function } x\colon V^{(\mathbb{B})} \longrightarrow \mathbb{B}$$

$$[\![x \,\varepsilon\, y]\!] = \bigvee_{u \in \mathrm{dom}(y)} [\![x \equiv u]\!] \wedge y(u)$$

$$[\![x \equiv y]\!] = \Big(\bigwedge_{u \in \mathrm{dom}(y)} \neg x(u) \vee [\![x \,\varepsilon\, y]\!]\Big) \wedge \Big(\bigwedge_{v \in \mathrm{dom}(y)} \neg y(v) \vee [\![v \,\varepsilon\, x]\!]\Big).$$

Here, this Boolean structure will be called $M(\mathbb{B})$. Define recursively the map $\check{}$ that takes sets $x$ to their Boolean analog,

$$\check{x} = \{\langle \check{y}, \mathbb{1}\rangle \mid y \in x\} \in V^{(\mathbb{B})}.$$

Define the Boolean pair and ordered pair:

$$\check{\{x,y\}} = \{<x,\mathbb{1}>,<y,\mathbb{1}>\}$$

$$\check{<x,y>} = \check{\{}\check{\{x\}},\check{\{x,y\}}\}.$$

For nonempty $N \subseteq V^{(\mathbb{B})}$, $N$ will be called dom-transitive iff

$$\forall xy(x \in dom(y) \ \& \ y \in N \to x \in N).$$

A dom-transitive $N$ determines a Boolean substructure $N$ of $M(\mathbb{B})$. Then $N$ is an interpretation of the language $L\varepsilon$. The axioms of extensionality and equality are Boolean valid for $N$, by the argument found in [22] for $V^{(\mathbb{B})}$.

For a complete subalgebra $\mathbb{C}$ of $\mathbb{B}$, $M(\mathbb{C})$ is a Boolean substructure of $M(\mathbb{B})$:

$$V^{(\mathbb{C})} \subseteq V^{(\mathbb{B})};$$

$$[\![x \varepsilon y]\!]^{(\mathbb{C})} = [\![x \varepsilon y]\!] \quad \text{for } x,y \in V^{(\mathbb{C})};$$

and

$$[\![x \equiv y]\!]^{(\mathbb{C})} = [\![x \equiv y]\!] \quad \text{for } x,y \in V^{(\mathbb{C})}.$$

(The superscript $(\mathbb{C})$ denotes the value for the structure $M(\mathbb{C})$.)

A Boolean isomorphism $\tau: \mathbb{B}_1 \to \mathbb{B}_2$ determines a 1-1 onto map $\underline{\tau}: V^{(\mathbb{B}_1)} \to V^{(\mathbb{B}_2)}$ by

$$\underline{\tau}(x) = \{<\underline{\tau}(y),\tau(x(y))> \mid y \in dom(x)\}.$$

Then function $\underline{\tau}$ is primitive recursive in the function $\tau$; that is, if $\tau$ is a (partial) primitive recursive function (with parameters), so is $\underline{\tau}$. Together, $\tau$ and $\underline{\tau}$ form an isomorphism of Boolean structures:

$$\tau[\![x \varepsilon y]\!] = [\![\underline{\tau}x \varepsilon \underline{\tau}y]\!],$$

$$\tau[\![x \equiv y]\!] = [\![\underline{\tau}x \equiv \underline{\tau}y]\!].$$

For another isomorphism $\sigma: \mathbb{B}_2 \to \mathbb{B}_3$, $\underline{\sigma\tau} = \underline{\sigma}\,\underline{\tau}$. If $\tau$ is identity, so is $\underline{\tau}$. For complete subalgebra $\mathbb{B}_0$ of $\mathbb{B}_1$, $\underline{\tau \restriction \mathbb{B}_0} = \underline{\tau} \restriction V^{(\mathbb{B}_0)}$. We generally write $\tau$ for both $\tau$ and $\underline{\tau}$.

3.4. Let $M$ be an admissible set. Let $\langle \mathbb{P}, \leq \rangle \in M$ where $\langle \mathbb{P}, \leq \rangle$ is a dense substructure of some complete Boolean algebra. Let $\mathcal{B}$ be the complete Boolean algebra $\{b \mid b \subseteq \mathbb{P} \ \& \ b = \neg\neg b\}$ ordered by $\subseteq$, as above. Then, for $\mathcal{B}$, $\leq$ is $\subseteq$; $\wedge$ is intersection $\cap$; and complement is the $\neg$ defined above on $\mathcal{P}(\mathbb{P})$. For this $\mathcal{B}$, the following relations and functions are primitive recursive with parameter $\langle \mathbb{P}, \leq \rangle$:

$$\mathcal{B},\ \leq,\ \wedge,\ \neg,\ V^{(\mathcal{B})},\ [\![\varepsilon]\!],\ [\![\equiv]\!],\ \text{and}\ \check{}\ .\quad \text{(Note 0.9 (3).)}$$

For example, the definition in 3.3 of $[\![\varepsilon]\!]$ and $[\![\equiv]\!]$ uses simultaneous double recursion, but it can be reduced to course of values recursion on $x$:

$$F(x,t) =$$

$$\{\langle y, b\wedge c\rangle \mid y \in t \ \& \ b = \bigwedge_{u\in \mathrm{dom}(x)} (\neg x(u) \vee \bigvee_{v\in \mathrm{dom}(y)} (F(u,t)(v) \wedge y(v)))$$

$$\& \ c = \bigwedge_{v\in \mathrm{dom}(y)} (\neg y(v) \vee \bigvee_{u\in \mathrm{dom}(x)} (F(u,t)(v) \wedge x(u)))\}$$

$$[\![x \equiv y]\!] = F(x, TC(\{y\}))(y)$$

$$[\![x \ \varepsilon\ y]\!] = \bigvee_{u\in \mathrm{dom}(y)} [\![x \equiv u]\!] \wedge y(u).$$

Thus, the restrictions to $M$ of these relations and functions are $M$-$\Delta_1$; the corresponding definitions $M$-define:

$$\mathcal{B} \cap M,\ \leq \cap M,\ \wedge \restriction (\mathcal{P}(\mathcal{B}) \cap M),\ \neg \restriction (\mathcal{B} \cap M),\ V^{(\mathcal{B})} \cap M,\ [\![\varepsilon]\!] \restriction M,$$

$$[\![\equiv]\!] \restriction M,\ \check{} \restriction M.$$

Put $M(\mathbb{P}) = V^{(\mathcal{B})} \cap M$.

Let $N$ be any dom-transitive $M$-$\Delta_1$ subset of $M(\mathbb{P})$. ($M(\mathbb{P})$ is such a N.) As mentioned in 3.3, N is an interpretation of the language $L\varepsilon$. Thus in 3.2 we defined the Boolean value $[\![\varphi]\!]$ for N and the language $L\varepsilon$.

We write $p \Vdash \varphi(\vec{x})$ for $p \in [\![\varphi(\vec{x})]\!]$. Then $p \Vdash \varphi(\vec{x})$ is M-definable, as a relation of $p$, $\vec{x}$, for each $\varphi \in L\varepsilon$. This is suggested by the following:

| | |
|---|---|
| $\varphi$ | $p \Vdash \varphi$ |
| $x \varepsilon y$ | $p \in [\![x \varepsilon y]\!]$ |
| $x \equiv y$ | $p \in [\![x \equiv y]\!]$ |
| $\neg\varphi$ | $\forall q \leq p \neg(q \Vdash \varphi)$ |
| $\varphi \& \psi$ | $(p \Vdash \varphi) \& (p \Vdash \psi)$ |
| $\forall v\varphi(v)$ | $\forall x(x \in N \rightarrow p \Vdash \varphi(x))$. |

The following is also useful.

| | |
|---|---|
| $\exists v\varphi(v)$ | $\forall q \leq p \exists r \leq q \exists x(x \in N \& r \Vdash \varphi(x))$ |
| $\exists v \varepsilon y\ \varphi(v)$ | $\forall q \leq p \exists r \leq q \exists x \in \mathrm{dom}(y)(r \in y(x) \& r \Vdash \varphi(x))$ |
| $\forall v \varepsilon y\ \varphi(v)$ | $\forall x \in \mathrm{dom}(y) \forall q \leq p(q \in y(x) \rightarrow q \Vdash \varphi(x))$ or $\forall x \in \mathrm{dom}(y)(p \Vdash (x \varepsilon y \rightarrow \varphi(x)))$ |
| $\varphi \rightarrow \psi$ | $\forall q \leq p((q \Vdash \varphi) \rightarrow (q \Vdash \psi))$. |

Because of our restrictions on N, if $\varphi$ is gen$\Sigma_n$, then $p \Vdash \varphi(\vec{x})$ is M-gen$\Sigma_n$ for $n > 0$. Similarly, if $\varphi$ is gen$\Pi_n$, then $p \Vdash \varphi(\vec{x})$ is M-gen$\Pi_n$ for $n > 0$. In particular, if $\varphi$ is $\Sigma_0$, then $p \Vdash \varphi(\vec{x})$ is M-$\Delta_1$ (actually, it is the restriction of a primitive recursive relation to N).

Except for $\Sigma_0$-reflection, [22] essentially shows that $M(\mathbb{P})$ is a Boolean model of the axioms of admissibility. The $\Sigma_0$-reflection principle also holds for $M(\mathbb{P})$; compare the more complicated Lemma 4.10 below. (If $M$ is a model of the power set axiom, then $M(\mathbb{P})$ is a Boolean model for that axiom; also, gen$\Delta_n$-separation, gen$\Sigma_n$-separation, and $\Sigma_n$-reflection are each a property inheritable by $M(\mathbb{P})$ from $M$.)

(Instead of assuming that $M$ is admissible, we could have assumed that $M$ was Prim-closed, transitive and nonempty. Then the above discussion still holds, provided we replace $\Sigma_0$-reflection by the property of being Prim-closed.)

# SECTION 4

## SOME BOOLEAN ADMISSIBLE SETS

Let $\aleph$ be an uncountable cardinal. Let $\mathbb{P} = \{P \mid P$ is a finite partial map $\aleph \times \omega \longrightarrow 2\}$. The letter $P$ will always denote a member of this $\mathbb{P}$.

The partial ordering $\langle \mathbb{P}, \supseteq \rangle$ is a dense substructure of an algebraically unique complete Boolean algebra $\langle \mathbb{B}, \leq \rangle$. Indeed, we now show

$$\forall P P_1 (P \not\supseteq P_1 \longrightarrow \exists P_2 \supseteq P \ \forall P_3 \supseteq P_2 (P_3 \not\supseteq P_1)).$$

Assume $P \not\supseteq P_1$. Then there exists some $\langle \alpha, n, k \rangle \in P_1 - P$. Put $P_2 = P \cup \{\langle \alpha, n, 1-k \rangle\} \in \mathbb{P}$. Consider any $P_3 \supseteq P_2$. Then $\langle \alpha, n, k \rangle \notin P_3$, since $P_3$ is a function and $\langle \alpha, n, 1-k \rangle \in P_3$. Thus,

$$\langle \alpha, n, k \rangle \in P_1 - P_3 \quad \text{and} \quad P_3 \not\supseteq P_1.$$

For each $u \subseteq \aleph$, define

$$\mathbb{P}u = \{P \mid P \text{ is a finite partial map } u \times \omega \longrightarrow 2\}.$$

Then $\mathbb{P}u \subseteq \mathbb{P}$. For each $P \in \mathbb{P}$, define $Pu = P \restriction u \times \omega \in \mathbb{P}u$. Then $P \leq P_1$ iff both $Pu \leq P_1 u$ and $P(\aleph - u) \leq P_1(\aleph - u)$. Also, $P = Pu \cup P(\aleph - u) = Pu \wedge P(\aleph - u)$.

4.1. LEMMA. *Put* $\mathbb{B}u = \{b \in \mathbb{B} \mid \forall P (P \leq b \longrightarrow Pu \leq b)\}$. *Then* $\mathbb{P}u$ *is a dense substructure of the complete subalgebra* $\mathbb{B}u$ *of* $\mathbb{B}$.

PROOF: $\mathbb{P}u \subseteq \mathbb{B}u$: if $P_1 \leq P \in \mathbb{P}u$, then $P_1 u \leq P$.

$\mathbb{P}u$ is dense in the set $\mathbb{B}u$: If $0 \neq b \in \mathbb{B}u$, then some $P \leq b$, since $\mathbb{P}$ is dense in $\mathbb{B}$. But then $b \geq Pu \in \mathbb{P}u$.

$\mathbb{B}u$ is closed under $\wedge$: Assume $X \subseteq \mathbb{B}u$. Then for each $P$

$$P \leq \wedge X \Longrightarrow \forall b \in X (P \leq b) \Longrightarrow \forall b \in X (Pu \leq b) \Longrightarrow Pu \leq \wedge X.$$

$\mathbb{B}u$ is closed under $\neg$: Assume $b \in \mathbb{B}u$. Then assume $P \leq \neg b$. We show $Pu \leq \neg b$ by showing $\forall P_1 \leq Pu(P_1 \nleq b)$. Assume $P_1 \leq Pu$. Consider $P_2 = P_1u \cup P(\aleph - u) \leq Pu \cup P(\aleph - u) = P$. Since $P \leq \neg b$, $\forall P_2 \leq P(P_2 \nleq b)$. Thus, $P_2 \nleq b$. If $P_1 \leq b$, then $b \geq P_1u \geq P_2$, a contradiction; thus, $P_1 \nleq b$.

4.2. LEMMA. Symmetry.

1) *A 1-1 onto map* $g: u \to v$ *determines an isomorphism* $g_{\mathbb{P}}: \mathbb{P}u \to \mathbb{P}v$ *by*

$$\langle\alpha,n,k\rangle \longmapsto \langle g(\alpha),n,k\rangle,$$

*namely*

$$g_{\mathbb{P}}(\{\langle\alpha_1,n_1,k_1\rangle,\ldots,\langle\alpha_m,n_m,k_m\rangle\}) = \{\langle g(\alpha_1),n_1,k_1\rangle,\ldots,\langle g(\alpha_m),n_m,k_m\rangle\}.$$

*For the unique isomorphism* $g_{\mathbb{B}}: \mathbb{B}u \to \mathbb{B}v$ *extending* $g_{\mathbb{P}}$:

$$(g \upharpoonright w)_{\mathbb{B}} = g_{\mathbb{B}} \upharpoonright \mathbb{B}w \quad \textit{for} \quad w \subseteq u;$$

$$(g_1 g)_{\mathbb{B}} = g_{1\mathbb{B}} g_{\mathbb{B}} \quad \textit{for 1-1 onto} \quad g_1: v \to w;$$

*if* $g$ *is identity, then* $g_{\mathbb{B}}$ *is identity.*

2) *A finite partial map* $p: \aleph \times \omega \to 2$ *determines an isomorphism* $p_u: \mathbb{P}u \to \mathbb{P}u$ *by*

$$\langle\alpha,n,1\rangle \longmapsto \langle\alpha,n,p(\alpha,n)\rangle \qquad \textit{for} \quad \langle\alpha,n\rangle \in \mathrm{dom}(p)$$

$$\langle\alpha,n,0\rangle \longmapsto \langle\alpha,n,1-p(\alpha,n)\rangle \qquad \textit{for} \quad \langle\alpha,n\rangle \in \mathrm{dom}(p)$$

$$\langle\alpha,n,k\rangle \longmapsto \langle\alpha,n,k\rangle \qquad \textit{for} \quad \langle\alpha,n\rangle \notin \mathrm{dom}(p).$$

*For the unique isomorphism* $p_u: \mathbb{B}u \to \mathbb{B}u$ *extending* $p_u: \mathbb{P}u \to \mathbb{P}u$:

$$p_w = p_u \upharpoonright \mathbb{B}w \quad \textit{for} \quad w \subseteq u;$$

*if* $p``u\times\omega \subseteq \{1\}$, *then* $p_u$ *is the identity.*

3) *Consider* $\gamma \in \omega$, *nonempty* 1-1 *finite sequence* $s: m \to \omega$, *and* 1-1 *onto* $h: \omega\text{-rng}(s) \to \omega - \{\gamma\}$. *These determine an isomorphism* $\underline{\gamma sh}: \mathbb{P}\omega \to \mathbb{P}\omega$ *by*

$$\langle s(r),n,k\rangle \longmapsto \langle \gamma, mn+r, k\rangle \qquad \textit{for} \quad r \in m$$

$$\langle \alpha,n,k\rangle \longmapsto \langle h(\alpha),n,k\rangle \qquad \textit{for} \quad \alpha \in \text{dom}(h).$$

*If* $z \subseteq \text{dom}(h)$ *and* $h \upharpoonright z$ *is identity, then* $\underline{\gamma sh} \upharpoonright \mathbb{P}z$ *is identity, and its extension* $\underline{\gamma sh} \upharpoonright \mathbb{B}z$ *is also the identity.*

4) *The above* (1,2,3) *determine isomorphisms of the respective Boolean structures* $V^{(\mathbb{B}u)}$, *etc. We call these isomorphisms* g, p, *and* $\underline{\gamma sh}$ *respectively. We omit the subscripts* $\mathbb{P}$, $\mathbb{B}$, *and* u. *The* u *of symmetry* (2) *is to be selected to fit the context.*

PROOF: The proof is left to the reader. For example, $(g \upharpoonright w)_{\mathbb{B}} = g_{\mathbb{B}} \upharpoonright \mathbb{B}w$ follows from the fact that $(g \upharpoonright w)_{\mathbb{B}}$ and $g_{\mathbb{B}} \upharpoonright \mathbb{B}w$ are each the unique isomorphism extending $(g \upharpoonright w)_{\mathbb{P}} = g_{\mathbb{P}} \upharpoonright \mathbb{P}w$. QED

Since $\mathbb{P}\omega$ is dense in $\mathbb{B}\omega$, we defined in 3.1 an identification $\mu: \mathbb{B}\omega \to \mathcal{B}\omega$ where $\mathcal{B}\omega = \{b \mid b \subseteq \mathbb{P}\omega \text{ and } b = \neg\neg b\}$ (here, $\neg$ is that operation defined on $\mathcal{P}(\mathbb{P}\omega)$). For each $z \subseteq \omega$, this $\mu$ identifies $\mathbb{B}z$ with $\mathcal{B}^*z = \mu``\mathbb{B}z = \{b \in \mathcal{B}\omega \mid \forall P \in b(Pz \in b)\}$, since

$$b \in \mu``\mathbb{B}z \leftrightarrow \mu^{-1}b \in \mathbb{B}z \leftrightarrow \forall P(P \le \mu^{-1}b \to Pz \le \mu^{-1}b) \leftrightarrow \forall P(P \in b \to Pz \in b).$$

For each $z \subseteq \omega$, the relation $V^{(\mathcal{B}^*z)}$ is primitive recursive in parameters z and $\mathbb{P}\omega$:

$$b \in \mathcal{B}^*z \leftrightarrow b \in \mathcal{B}\omega \ \& \ \forall P \in b(Pz \in b)$$

$$x \in V^{(\mathcal{B}^*z)} \leftrightarrow x \in V^{(\mathcal{B}\omega)} \ \& \ \forall y \in \text{dom}(x)(y \in V^{(\mathcal{B}^*z)} \ \& \ x(y) \in \mathcal{B}^*z).$$

Let admissible set M be given such that $\omega \in M$. Then $\mathbb{P}\omega \in M$. We now begin to define Nu (which depends on M) for each $u \subseteq \aleph$.

The $Mz$ of Lemma 1.2 will be $N\omega \cup z$ "reduced" to a transitive set.

4.3. <u>Definition</u>. For each finite $z \subseteq \omega$, put $N^*z = V^{(\mathcal{B}^*z)} \cap M$ and $Nz = \mu^{-1}``N^*z$. Since $V^{(\mathcal{B}^*z)}$ is primitive recursive with parameters in $M$, $N^*z$ is $M$-$\Delta_1$. The restriction to $M$ of the above definition of $V^{(\mathcal{B}^*z)}$ yields $x \in N^*z \leftrightarrow x \in M(\mathbb{P}\omega)$ & $\forall y \in \mathrm{dom}(x)(y \in N^*z$ & $x(y) \in \mathcal{B}^*z)$. Note that $z \subseteq z_1$ implies $N^*z \subseteq N^*z_1$ and $Nz \subseteq Nz_1$. Also, $Nz \subseteq V^{(\mathbb{B}z)}$.

4.4. LEMMA.

1) *For finite* $z, z_1 \subseteq \omega$, *maps* $g$, $g_1$ *of symmetry* (1), $z \subseteq \mathrm{dom}(g)$, $z_1 \subseteq \mathrm{dom}(g_1)$, *and* $g``z = g_1``z_1$:

$$g``Nz = g_1``Nz_1.$$

2) *For finite* $z \subseteq \omega$, $g$ *of symmetry* (1), $z \subseteq \mathrm{dom}(g)$, *and* $p$ *of symmetry* (2):

$$pg``Nz = g``Nz.$$

3) *For finite* $z \subseteq \omega$ *and* $\gamma$, $s$, $h$ *of symmetry* (3):

$$\underline{\gamma sh}``N(z \cup \mathrm{rng}(s)) = N(\{\gamma\} \cup h``z).$$

PROOF: (1) Put finite $h = g_1^{-1}g \restriction z \in M$. Now $h\colon \mathbb{P}z \longrightarrow \mathbb{P}z_1$ is a member of $M$. Further, $h^*\colon \mathcal{B}^*z \longrightarrow \mathcal{B}^*z_1$ was defined in 3.1 (where $\mathbb{B}$ there becomes $\mathbb{B}\omega$ here, etc.) and it was primitive recursive in $h$ and $\mathbb{P}\omega$. Thus, $h^*\colon V^{(\mathcal{B}^*z)} \longrightarrow V^{(\mathcal{B}^*z_1)}$ maps $M$ into $M$. Thus $h^*``N^*z \subseteq N^*z_1$. Similarly, $h^{-1*}``N^*z_1 \subseteq N^*z$. Combining, $h^*``N^*z = N^*z_1$. Since $h = \mu^{-1}h^*\mu$, $h``Nz = Nz_1$. Thus $g``Nz = (g_1 g_1^{-1} g \restriction z)``Nz = g_1 h``Nz = g_1``Nz_1$.

(2) Consider the finite $p_1$ for symmetry (2) defined by

$$p_1(\alpha, n) = \begin{cases} p(g(\alpha), n) & \text{if defined} \\ \text{undefined otherwise.} \end{cases}$$

Then $g^{-1}pg \restriction \mathbb{P}u$ is the same as $p_1 : \mathbb{P}u \to \mathbb{P}u$. Further, $p_1 \restriction \mathbb{P}z \in M$. Then $(p_1 \restriction \mathbb{P}z)^* : V^{(\mathcal{B}^* z)} \to V^{(\mathcal{B}^* z)}$ maps M into M, as does its inverse; $(p_1 \restriction \mathbb{P}z)^{*}``N^*z = N^*z$. Therefore, $p_1``Nz = Nz$ and

$$pg``Nz = gg^{-1}pg``Nz = gp_1``Nz = g``Nz.$$

(3) Consider $\underline{\gamma sh} : \mathbb{P}\omega \to \mathbb{P}\omega$. Then $\underline{\gamma sh} \restriction \mathbb{P}z \cup rng(s)$ is an isomorphism onto $\mathbb{P}\{\gamma\} \cup h``z$. Further, $\underline{\gamma sh} \restriction \mathbb{P}z \cup rng(s)$ is a member of M. Thus, $(\underline{\gamma sh} \restriction \mathbb{P}z \cup rng(s))^*$ maps M into M, as does its inverse. Thus, $(\underline{\gamma sh} \restriction \mathbb{P}z \cup rng(s))^{*}``N^*(z \cup rng(s)) = N^*(\{\gamma\} \cup h``z)$. QED

Consider any finite $u \subseteq \omega$. For any finite $z \subseteq \omega$ and 1-1 onto $g: z \to g``z$, $g``Nz = Ng``z \subseteq Nu$ when $g``z \subseteq u$. Also, for the identity $h: u \to u$, $Nu = h``Nu$. Thus, we can extend the definition of Nu as follows:

4.5 Definition. For $u \subseteq \aleph$, Nu is defined to be

$$Nu = \bigcup\{g``Nz \mid \text{finite } z \subseteq \omega \text{ and 1-1 } g: z \to u\}.$$

Then $Nu \subseteq V^{(\mathbb{B}u)}$.

If $u \subseteq v \subseteq \aleph$, then $Nu \subseteq Nv$.

If $u, v \subseteq \aleph$ and if $h: u \to v$ is 1-1 onto, then

$$\begin{aligned} h``Nu &= \bigcup\{hg``Nz \mid \text{finite } z \subseteq \omega \text{ and 1-1 } g: z \to u\} \\ &= \bigcup\{g_1``Nz \mid \text{finite } z \subseteq \omega \text{ and 1-1 } g_1: z \to v\} \\ &= Nv. \end{aligned}$$

In particular, for finite $z \subseteq \omega$ and 1-1 $g: z \to g``z$, $g``Nz = Ng``z$. Thus,

$$\begin{aligned} Nu &= \bigcup\{Ng``z \mid \text{finite } z \subseteq \omega \text{ and 1-1 } g: z \to u\} \\ &= \bigcup\{Nz \mid \text{finite } z \subseteq u\}. \end{aligned}$$

Then $\mu^{-1}``N\omega$ is $\bigcup\{N^*z \mid \text{finite } z \subseteq \omega\}$. Call it $N^*\omega$. We will show later that $N^*\omega$ is a Boolean model of the admissibility axioms, so that $N\omega$ is also. For the present, we note that $N^*\omega$ is a dom-transitive subset of $M(\mathbb{P}\omega)$; it is $M$-$\Delta_1$ since

$$x \in N^*\omega \leftrightarrow \exists z(\text{finite } z \subseteq \omega \;\&\; x \in N^*z).$$

4.6. LEMMA.

1) *For* $g: u \to v$ *of symmetry* (1), $g \restriction Nu$ *is an isomorphism of the Boolean substructures* $Nu$ *and* $Nv$ *of* $V^{(\mathbb{B})}$.

2) *For* $p$ *and* $u$ *of symmetry* (2), $p \restriction Nu$ *is an automorphism of the Boolean substructure* $Nu$.

3) *For* $\gamma$, $s$, $h$ *of symmetry* (3), $\underline{\gamma sh} \restriction N\omega$ *is an automorphism of* $N\omega$.

PROOF: (1) We have shown $g``Nu = Nv$. As mentioned in 3.2, this suffices.

(2) Lemma 4.4 (2) implies $p``Nu = Nu$:

$$\begin{aligned} p``Nu &= \bigcup\{pg``Nz \mid \text{finite } z \subseteq \omega \text{ and } 1\text{-}1 \;\; g: z \to u\} \\ &= \bigcup\{g``Nz \mid \text{finite } z \subseteq \omega \text{ and } 1\text{-}1 \;\; g: z \to u\} \\ &= Nu. \end{aligned}$$

(3) Lemma 4.4 (3) implies $\underline{\gamma sh}``N\omega = N\omega$. QED

Define $[\![\varphi(\vec{x})]\!]_u$ to be the Boolean value for the Boolean substructure $\langle Nu, [\![\epsilon]\!], [\![\equiv]\!]\rangle$ of $V^{(\mathbb{B})}$. Define $P \Vdash_u \varphi(\vec{x})$ to mean $P \le [\![\varphi(\vec{x})]\!]_u$.

4.7. LEMMA. *If* $\vec{x} \in Nz$ *and* $z \subseteq u$, *then* $[\![\varphi(\vec{x})]\!]_u \in \mathbb{B}z$.

PROOF: It suffices to show that $P \Vdash_u \varphi(\vec{x})$ implies $Pz \Vdash_u \varphi(\vec{x})$. Assume $P' \le Pz$. For symmetry (2), find $p: \mathrm{dom}(P) \cap \mathrm{dom}(P') \to 2$

such that $(pP)\restriction \mathrm{dom}(P') \subseteq P'$ (i.e., define $p$ by

$$p(\alpha,n) = 1 \quad \text{if } P(\alpha,n) = P'(\alpha,n) \text{ defined}$$

$$p(\alpha,n) = 0 \quad \text{if } P(\alpha,n) \neq P'(\alpha,n), \text{ both defined}$$

$$p(\alpha,n) \quad \text{undefined otherwise}).$$

Then $P' \geq P' \cup pP \Vdash_u \varphi(p\vec{x})$. Since $P'$ extends $Pz$, $p$ is the identity on $\mathbb{P}z$ and on $V^{(\mathbb{B}z)}$. Thus, $p\vec{x} = \vec{x}$. This shows that

$$\forall P' \leq Pz\, \exists P'' \leq P'(P'' \Vdash_u \varphi(\vec{x})).$$

By the (extended) contrapositive of the condition for being a dense set, $Pz \Vdash_u \varphi(\vec{x})$.

4.8. LEMMA. *If* $\vec{x} \in Nz$, *finite* $z \subseteq u \cap v$, *and* $u \simeq v$, *then* $[\![\varphi(\vec{x})]\!]_u = [\![\varphi(\vec{x})]\!]_v \in \mathbb{B}z$.

PROOF: By the previous lemma, both values are in $\mathbb{B}z$. For symmetry (1), find 1-1 onto $g: u \longrightarrow v$ such that $g\restriction z$ is identity. Then $g\restriction \mathbb{B}z$ is identity; $[\![\varphi(\vec{x})]\!]_u = g([\![\varphi(\vec{x})]\!]_u) = [\![\varphi(g\vec{x})]\!]_v = [\![\varphi(\vec{x})]\!]_v$.

4.9. THEOREM. $N\omega$ *is a Boolean model of the axioms of admissibility. It is dom-transitive, as is every* $Nu$.

PROOF: The dom-transitivity follows directly from the definitions.

We now prove that $N^*\omega$ is a Boolean model of the axioms of admissibility. Note that $N^*\omega$ is an $N$ of 3.4, i.e., it is dom-transitive and $M$-$\Delta_1$. We use the notation there for the structure $N = N^*\omega$. Then for gen$\Sigma_{n+1}$ formula $\varphi$, $P \Vdash \varphi(\vec{x})$ is $M$-gen$\Sigma_{n+1}$. For the purpose of this proof, $P$ denotes a member of $\mathbb{P}\omega$.

The following are in $N^*\omega$ when $x$, $y$, $\vec{x}$ are in $N^*\omega$ and $\varphi$ is $\Sigma_0$:

$$\{\langle x,\mathbb{1}\rangle,\langle y,\mathbb{1}\rangle\}$$

$$\{\langle y, [\![\exists v_1 \in x(y \in v_1)]\!]\rangle \mid \exists z \in \mathrm{dom}(x)(y \in \mathrm{dom}(z))\}$$

$$\{\langle y, [\![y \in x \ \&\ \varphi(\vec{x},y)]\!]\rangle \mid y \in \mathrm{dom}(x)\}.$$

This is because $M$ is Prim-closed and the above are values of primitive recursive functions at $x, y, \vec{x}, \varphi$; and these values are in some $V^{(\mathcal{B}^* z)}$ when $x, y, \vec{x}$ are, using Lemma 4.7 (and $\mu: \mathbb{B}\omega \to \mathcal{B}\omega$). Given that these values are in $N^*\omega$, the verification of the axioms of pair, union, and $\Sigma_0$-separation proceeds as in [22]. (For $\Sigma_0$-separation, see the proof of [22] for the comprehension schema; it is considered obvious that $\{\langle x,\mathbb{1}\rangle,\langle y,\mathbb{1}\rangle\}$ is the "pair.")

<u>Digression</u>. If $M$ satisfies the schema of gen$\Sigma_n$-separation, then $N^*\omega$ does also: Let $\varphi(\vec{v},v_1)$ be gen$\Sigma_n$. Let $x,\vec{x} \in N^*\omega$. We will find $x_2$ such that $[\![\forall v_1(v_1 \in x_2 \leftrightarrow v_1 \in x \ \&\ \varphi(\vec{x},v_1))]\!] = \mathbb{1}$. First,

$$x_1 \in \mathrm{dom}(x) \ \&\ P \Vdash x_1 \in x \ \&\ \varphi(\vec{x},x_1)$$

is $M$-gen$\Sigma_n$ (or $M$-$\Delta_1$ if $n = 0$). Put

$$w = \{\langle x_1,P\rangle \mid x_1 \in \mathrm{dom}(x) \ \&\ P \Vdash x_1 \in x \ \&\ \varphi(\vec{x},x_1)\}.$$

By gen$\Sigma_n$-separation (or $\Delta_1$-separation), $w \in M$, since $\mathrm{dom}(x) \times \mathbb{P}\omega \in M$. Put

$$x_2 = \{\langle x_1,b\rangle \mid x_1 \in \mathrm{dom}(x) \ \&\ b = \{P \mid \langle x_1,P\rangle \in w\}\}.$$

Then $x_2 \in M$, since $M$ is Prim-closed. Further,

$$(*) \qquad x_2 = \{\langle x_1, [\![x_1 \in x \ \&\ \varphi(\vec{x},x_1)]\!]\rangle \mid x_1 \in \mathrm{dom}(x)\}.$$

Find finite $z \subseteq \omega$ such that $x,\vec{x} \in N^*z$. By Lemma 4.7, $[\![x_1 \in x \ \&\ \varphi(\vec{x},x_1)]\!] \in \mathcal{B}^*z$ for all $x_1 \in \mathrm{dom}(x) \subseteq N^*z$. Thus, $x_2 \in N^*z \subseteq N^*\omega$. The proof that $x_2$ is as desired is the proof of the comprehension schema given in [22], using equation (*). (The inheritance of gen$\Delta_n$-separation is slightly harder to show.)

4.10. LEMMA. *If* M *satisfies the schema of* $\Sigma_n$*-reflection, then* $N^*\omega$ *does also. In particular,* $N^*\omega$ *satisfies the schema of* $\Sigma_0$*-reflection.*

PROOF: Let $\varphi(\vec{v},v_0,v_1)$ be $\Sigma_n$. Let $x,\vec{x} \in N^*\omega$. Assume

$$P \Vdash \forall v_0 \varepsilon x \; \exists v_1 \varphi(\vec{x},v_0,v_1);$$

we will show that there is $x_2$ such that

$$P \Vdash \forall v_0 \;\varepsilon\; x \exists v_1 \;\varepsilon\; x_2 \varphi(\vec{x},v_0,v_1).$$

Find finite $z \subseteq \omega$ such that $x,\vec{x} \in N^*z$. By Lemma 4.7, we can assume $P \in \mathbb{P}z$. Find $\gamma \in \omega - z$.

For each $x_0 \in \mathrm{dom}(x)$:

$$P \Vdash x_0 \;\varepsilon\; x \rightarrow \exists v_1 \varphi(\vec{x},x_0,v_1);$$

$$P \Vdash \exists v_1 (x_0 \;\varepsilon\; x \rightarrow \varphi(\vec{x},x_0,v_1));$$

$$\forall P_0 \leq P \; \exists P_1 \leq P_0 \; \exists x_1 \in N^*\omega (P_1 \Vdash x_0 \;\varepsilon\; x \rightarrow \varphi(\vec{x},x_0,x_1)).$$

From this we show first that

$$\forall P_0 \leq P \; \exists P_3 \leq P_0 \quad \exists u \in N^*(z \cup \{\gamma\})(P_1 \Vdash x_0 \;\varepsilon\; x \rightarrow \varphi(\vec{x},x_0,u)).$$

Assume $P_0 \leq P$. Then, for some $P_1 \leq P_0$ and $x_1 \in N^*\omega$,

$$P_1 \Vdash x_0 \;\varepsilon\; x \rightarrow \varphi(\vec{x},x_0,x_1).$$

Find finite $z' \subseteq \omega$ such that $x_1 \in N^*z'$. We can assume $z'-z \neq 0$. List $z'-z$ by a 1-1 finite sequence $s$. Find 1-1 onto $h: \omega - \mathrm{rng}(s) \rightarrow \omega - \{\gamma\}$ such that $h \restriction z$ is identity. By symmetry (3), take $\tau = \underline{\gamma sh}^*$. Then

$$\tau P_1 \Vdash \tau x_0 \;\varepsilon\; \tau x \rightarrow \varphi(\tau\vec{x},\tau x_0,\tau x_1);$$

$\tau P_1 = P_1 z \cup P_2$ for some $P_2 \in \mathbb{P}\omega - z$; $\tau x = x$; $\tau\vec{x} = \vec{x}$; $\tau x_0 = x_0$; $\tau x_1 \in N^*(z \cup \{\gamma\})$. For symmetry (2), find $p: \mathrm{dom}(P_2) \cap \mathrm{dom}(P_0) \rightarrow 2$

such that $(pP_2)\restriction \mathrm{dom}(P_0) \subseteq P_0$. Note that $p\restriction\mathbb{B}^*z$ is identity. Then

$$P_1z \cup pP_2 = p(P_1z \cup P_2) \Vdash px_0 \in px \to \varphi(p\vec{x}, px_0, p\tau x_1);$$

$px = x$; $p\vec{x} = \vec{x}$; $px_0 = x_0$; $p\tau x_1 \in N^*(z \cup \{\gamma\})$. Put $P_3 = P_1z \cup pP_2 \cup P_0 \leq P_0$; put $u = p\tau x_1$.

We have shown

$$\forall s \in (\mathrm{dom}(x) \times \{P_0 \mid P_0 \leq P\})\exists u\exists P_3 \leq [s]_1(u \in N^*(z \cup \{\gamma\})$$

$$\& \ P_3 \Vdash [s]_0 \in x \to \varphi(\vec{x}, [s]_0, u)).$$

By $\Sigma_n$-reflection (or, if $n = 0$, $\Sigma_1$-reflection), find a restriction $w \in M$ for the quantifier "$\exists u$." By $\Delta_1$-separation, $w \cap N^*(z \cup \{\gamma\}) \in M$. Put $x_2 = (w \cap N^*(z \cup \{\gamma\})) \times \{\mathbb{1}\} \in N^*(z \cup \{\gamma\})$.

We now show $P \Vdash \forall v_0 \in x\exists v_1 \in x_2\varphi(\vec{x}, v_0, v_1)$. This follows from $x_0 \in \mathrm{dom}(x)(P \Vdash x_0 \in x \to \exists v_1 \in x_2\varphi(\vec{x}, x_0, v_1))$ or the stronger

$$\forall x_0 \in \mathrm{dom}(x)\forall P_0 \leq P\ \exists P_3 \leq P_0\ \exists x_1 \in \mathrm{dom}(x_2)(\mathbb{1} \Vdash x_1 \in x_2$$

$$\& \ P_3 \Vdash x_0 \in x \to \varphi(\vec{x}, x_0, x_1)).$$

Assume $x_0 \in \mathrm{dom}(x)$ and $P_0 \leq P$. By the choice of $x_2$, there are $u \in \mathrm{dom}(x_2)$ and $P_3 \leq P_0$ such that $[\![u \in x_2]\!] = \mathbb{1}$ and $P_3 \Vdash x_0 \in x \to \varphi(\vec{x}, x_0, u)$.

QED lemma and theorem.

The next lemma will be used for proving Lemma 1.2 (c).

4.11. LEMMA. *For each* $\alpha \in \aleph$, *put*

$$b_\alpha = \{\langle\check{n}, \{\langle\alpha, n, 1\rangle\}\rangle \mid n \in \omega\} \in V^{(\mathbb{B}\{\alpha\})}.$$

*Then* $b_\alpha \in N\{\alpha\}$. *For* $x \in V^{(\mathbb{B}(\aleph - \{\alpha\}))}$, $[\![b_\alpha \equiv x]\!] = \mathbb{0}$. *In particular, for* $\alpha \neq \beta \in \aleph$, $[\![b_\alpha = b_\beta]\!] = \mathbb{0}$.

PROOF: Since $\omega \in M$,

$$\mu b_0 \;=\; \{<\check{n},\{P \in \mathbb{P}\omega \mid <0,n,1> \in P\}> \mid n \in \omega\} \in M.$$

Thus, $\mu b_0 \in N^*\{0\}$ and $b_0 \in N\{0\}$. Find map $g$ for symmetry (1) such that $g(0) = \alpha$. Then $b_\alpha = gb_0 \in g``N\{0\} = N\{\alpha\}$.

If $[\![b_\alpha \equiv x]\!] \neq \mathbb{0}$, then some $P \leq [\![b_\alpha \equiv x]\!]$. Find $n \in \omega$ such that $<\alpha,n> \notin \mathrm{dom}(P)$. There is some $P' \leq P$ such that $P' \leq [\![\check{n} \in x]\!]$ or $P' \leq \neg[\![\check{n} \in x]\!]$. We just consider the case $P' \leq [\![\check{n} \in x]\!]$ (the other case is similar). Since $[\![\check{n} \in x]\!] \in \mathbb{B}(\aleph-\{\alpha\})$ and $V^{(\mathbb{B})}$ is a model of the extensionality axioms,

$$P \cup (P'(\aleph-\{\alpha\})) \cup \{<\alpha,n,0>\} \leq [\![b_\alpha \equiv x \;\&\; \check{n} \in x \;\&\; \neg\check{n} \in b_\alpha]\!] = \mathbb{0},$$

a contradiction. Thus, $[\![b_\alpha \equiv x]\!] = \mathbb{0}$.

The next lemma will be used for proving Lemma 1.2 (b). For 4.12, we write $P \Vdash \varphi$ or $P \in [\![\varphi]\!]$ for $P \in [\![\varphi]\!]^{(\mathcal{B}^*\omega)}$.

4.12. LEMMA. *If* $z \subseteq \omega$ *and* $z' \subseteq \omega$, *then there is a primitive recursive function* $F\colon V^{(\mathcal{B}^*z)} \times \mathbb{P}z \longrightarrow V^{(\mathcal{B}^*z\cap z')}$ *(in parameters) such that for all* $x \in V^{(\mathcal{B}^*z)}$:

$$\bigvee\{[\![x \equiv y]\!] \mid y \in V^{(\mathcal{B}^*z)}\} \;=\; \bigvee\{[\![x \equiv F(x,P)]\!] \mid P \in \mathbb{P}z\};$$

*if* $y \in V^{(\mathcal{B}^*z')}$ *and* $P \Vdash x \equiv y$, *then* $P \Vdash x \equiv F(x,Pz)$. *If* $x \in N^*z$, *then* $F(x,Pz) \in N^*z \cap z'$.

PROOF: Define the function $F$ by induction on $u \in \mathrm{dom}(x)$: $F(x,P)$ is a function with the domain $F``(\mathrm{dom}(x) \times \mathbb{P}z)$, and

$$F(x,P)(F(u,Q_1)) \;=\; \bigvee\{\mu Q \mid Q \in \mathbb{P}z \cap z' \;\&\; Q \leq Pz \cap z' \;\&\; Q \cup P \Vdash F(u,Q_1) \in x\}.$$

This $F$ is primitive recursive with parameters $z$, $z \cap z'$, $\mathbb{P}z$, and $\mathbb{P}z \cap z'$. Incidentally, its dependence on $z'$ is only a dependence on $z \cap z'$.

If $x \in N^* z$: $F(x,Pz) \in M \cap V^{(\mathcal{B}^* z \cap z')} = N^* z \cap z'$.

By induction on $u \in \mathrm{dom}(x)$, we prove: if $x \in V^{(\mathcal{B}^* z)}$, $y \in V^{(\mathcal{B}^* z')}$, and $P \Vdash x \equiv y$, then $P \Vdash x \equiv F(x,Pz)$.

$P \Vdash F(x,Pz) \subseteq x$: Suppose $P_0 \leq P$, $w \in \mathrm{dom}(F(x,Pz))$, and $P_0 \Vdash w \,\varepsilon\, F(x,Pz)$; we find an extension $P_2 \leq P_0$ such that $P_2 \Vdash w \,\varepsilon\, x$. By definition of $[\![\varepsilon]\!]$, there exist $P_1 \leq P_0$ and $w_1 \in \mathrm{dom}(F(x,Pz))$ such that $P_1 \Vdash w \equiv w_1$ and $P_1 \in F(x,Pz)(w_1)$. By defintiion of $F$, there exist $P_2 \leq P_1$ and $Q \in \mathbb{P}z \cap z'$ such that $P_2 \leq Q \leq Pz \cap z'$ and $Q \cup Pz \Vdash w_1 \,\varepsilon\, x$. Then $P_2 \leq P_1$, $Q \cup Pz$, so $P_2 \Vdash w \equiv w_1 \,\varepsilon\, x$ (i.e., the equality axiom for $V^{(\mathcal{B}^*\omega)}$ implies $P_2 \Vdash w \,\varepsilon\, x$).

$P \Vdash x \subseteq F(x,Pz)$: Suppose $P_0 \leq P$, $u \in \mathrm{dom}(x)$, and $P_0 \Vdash u \,\varepsilon\, x$; we find an extension $P_1 \leq P_0$ such that $P_1 \Vdash u \,\varepsilon\, F(x,Pz)$. Since $P_0 \Vdash u \,\varepsilon\, y$, there exist $P_1 \leq P_0$ and $v \in \mathrm{dom}(y) \subseteq V^{(\mathcal{B}^* z')}$ such that $P_1 \Vdash u \equiv v$. By induction hypothesis, $P_1 \Vdash u \equiv F(u,P_1 z)$. Thus, $P_1 \Vdash F(u,P_1 z) \equiv u \,\varepsilon\, x$. Now $(P_1 z \cap z') \cup Pz \Vdash F(u,P_1 z) \,\varepsilon\, x$:

$$P_1 \in [\![F(u,P_1 z) \,\varepsilon\, x \;\&\; x \equiv y]\!] \leq [\![F(u,P_1 z) \,\varepsilon\, y]\!] \in \mathcal{B}^* z'$$

$$P_1 z' \in [\![F(u,P_1 z) \,\varepsilon\, y]\!]$$

$$P_1 z' \cup P \in [\![F(u,P_1 z) \,\varepsilon\, y \;\&\; x \equiv y]\!] \leq [\![F(u,P_1 z) \,\varepsilon\, x]\!] \in \mathcal{B}^* z$$

$$(P_1 z \cap z') \cup Pz \in [\![F(u,P_1 z) \,\varepsilon\, x]\!].$$

By definition of $F$, $P_1 z \cap z' \in F(x,Pz)(F(u,P_1 z))$. Thus,

$$P_1 \in \mu(P_1 z \cap z') \wedge [\![F(u,P_1 z) \equiv u]\!] \leq F(x,Pz)(F(u,P_1 z)) \wedge [\![F(u,P_1 z) \equiv u]\!]$$

$$\leq [\![u \,\varepsilon\, F(x,Pz)]\!]. \quad \text{QED}$$

4.13. COROLLARY. *If* $x \in Nu$, $y \in V^{(\mathbb{B}v)}$, *and* $P \leq [\![x \equiv y]\!]$, *then there is* $w \in Nu \cap v$ *such that* $P \leq [\![x \equiv w]\!]$.

PROOF: For some finite $z \subseteq u$ and finite $z' \subseteq v$, $x \in Nz$ and

$y \in V^{(\mathbb{B}z')}$. We can assume $P \in \mathbb{P}z \cup z'$. Find 1-1 $g: z \cup z' \longrightarrow \omega$. Then $\mu x \in N^{*}g``z$, $\mu y \in V^{(\mathcal{B}^{*}g``z')}$, and $gP \Vdash \mu gx \equiv \mu gy$. By the last lemma, there is $t \in N^{*}((g``z) \cap (g``z'))$ such that $gP \Vdash \mu gx \equiv t$. Then $P \leq [\![x \equiv g^{-1}\mu^{-1}t]\!]$. Put $w = g^{-1}\mu^{-1}t \in Nz \cap z'$.

# SECTION 5

## SOME ADMISSIBLE SETS

Let M be a countable admissible set containing $\omega$. In 4.5, Nu was defined for each $u \subseteq \aleph_1$. We now restate Lemma 1.2.

5.1. LEMMA. *There exists an indexed set* $\{Mz \mid \text{finite } z \subseteq \aleph_1\}$ *of countable admissible sets for which:*

a) $z \subseteq z'$ *implies that* Mz *is an elementary submodel of* Mz'

b) $Mz \cap Mz' = Mz \cap z'$

c) *there is an indexed set* $\{a_\alpha \mid \alpha \in \aleph_1\}$ *such that:* $a_\alpha \in M\{\alpha\}$; $a_\alpha \neq a_\beta$ *for* $\alpha \neq \beta$

d) $Mz \supseteq M$ *and* $o(Mz) = o(M)$

e) *if* $N\omega$ *is a Boolean model of* $\varphi$, *then* M0 *is a model of* $\varphi$.

(Our particular indexing will have the peculiarity that $Mz = M(z-\omega)$.)

PROOF: We will find a function $G: \aleph_1 \times \omega \to 2$ such that, for each finite $z \subseteq \aleph_1$,

$$\{[\![\varphi]\!]_{\omega \cup z} \mid \exists P \subseteq G(P \leq [\![\varphi]\!]_{\omega \cup z})\}$$

is an ultrafilter $\mathcal{F}$ of the Boolean algebra of values of $L\varepsilon^+$ sentences with parameters in $N\omega \cup z$. Further, $\mathcal{F}$ will preserve some infinite meets corresponding to quantification and to the definition of $[\![\varepsilon]\!]$. Then Mz will be Mostowski's reduction of $N\omega \cup z/\mathcal{F}$ to a transitive set.

5.2. SUBLEMMA. *There is a function* $G: \aleph_1 \times \omega \to 2$ *such that for each finite* $z \subseteq \aleph_1$ *and for each* $L\varepsilon^+$ *sentence* $\varphi$ *with parameters in* Nz,

G *has the following property* $q = \langle z,\varphi\rangle$:

$\exists P \subseteq G \upharpoonright ((\omega \cup z) \times \omega)$:

(1) $P \leq [\![\varphi]\!]_{\omega\cup z}$ *or* $P \leq [\![\neg\varphi]\!]_{\omega\cup z}$;

(2) *if* $\varphi$ *is some* $\forall v\psi(v)$, *then* $P \leq [\![\neg\varphi]\!]_{\omega\cup z}$ *implies* $\exists u \in N\omega \cup z(P \leq [\![\neg\psi(u)]\!]_{\omega\cup z})$; *and*

(3) *if* $\varphi$ *is* $x \in y$, *then* $P \leq [\![x \in y]\!]$ *implies* $\exists u \in \mathrm{dom}(y)(P \leq [\![x \equiv u \;\&\; u \in y]\!])$.

PROOF: For a property $q$, we say $P$ gives property $q$ iff $P$ satisfies (1), (2), and (3). Corresponding to "$\exists u$" in clauses (2) and (3), a new parameter $u$ may arise; for (2), possibly $u \notin Nz$.

Given property $q = \langle z,\varphi(\vec{x})\rangle$ and 1-1 onto $g: \omega \cup z \to \omega \cup g``z$ (for symmetry (1)), define $gq$ to be the property $\langle g``z,\varphi(g\vec{x})\rangle$. For $p$ of symmetry (2), define $pq$ to be property $\langle z,\varphi(p\vec{x})\rangle$. If $P$ gives property $q$, then $gP$ gives property $gq$, and $pP$ gives property $pq$.

By induction on $\alpha \in [\omega,\aleph_1]$, we construct functions $G_\alpha: \alpha \times \omega \to 2$ such that the following $Q(\alpha)$ is true.

$Q(\alpha)$: For all $\beta \in [\omega,\alpha)$, $G_\beta \subseteq G_\alpha$: and for each property $q = \langle z,\varphi\rangle$ such that $z \subseteq \alpha$, there exists $P \subseteq G_\alpha \upharpoonright ((\omega \cup z) \times \omega)$ such that $P$ gives property $q$.

Induction stage $\alpha = \omega$: For $\alpha = \omega$, we construct what is usually called a complete generic sequence for $N\omega$ (note that $z \subseteq \omega$ implies $N\omega \cup z = N\omega$). Form a list $\langle q_n \mid n \in \omega\rangle$ of all the properties $q = \langle z,\varphi\rangle$ such that $z \subseteq \omega$. By induction on $n \in \omega$, find $P_n \in \mathbb{P}\omega$ such that: $P_n \supseteq \bigcup_{i<n} P_i$ and $P_n$ gives property $q_n$. (For example, when $q = \langle z,\forall v\psi(v)\rangle$, either $\bigcup_{i<n} P_i \leq [\![\forall v\psi(v)]\!]_{\omega\cup z}$, or there exists $u \in N\omega \cup z$ such that $\bigcup_{i<n} P_i \nleq [\![\psi(u)]\!]_{\omega\cup z}$. In the first case, put

$P_n = \bigcup_{i<n} P_i$. In the second case, there exists $P_n \supseteq \bigcup_{i<n} P_i$ such that $P_n \leq [\![\neg\psi(u)]\!]_{\omega\cup z}$.) Then $\bigcup_{i<\omega} P_i : \omega \times \omega \to 2$ is total (since for $\langle n,m \rangle \in \omega \times \omega$, some $P_i$ gives property $\langle \{n\}, \check{m} \varepsilon b_n \rangle$, i.e., either $P_i \leq [\![\check{m} \varepsilon b_n]\!] = \{\langle n,m,1 \rangle\}$ or $P_i \leq [\![\neg \check{m} \varepsilon b_n]\!] = \{\langle n,m,0 \rangle\}$). Put $G'_\omega = \bigcup_{i<\omega} P_i$. Then $Q(\omega)$ is true.

Induction stage $\alpha > \omega$: By induction hypothesis, for all $\beta \in [\omega,\alpha)$, $G_\beta$ has been constructed and $Q(\beta)$ is true. Either $\alpha$ is a limit ordinal or, for some $\beta \in [\omega,\alpha)$, $\alpha = \beta+1$.

If $\alpha$ is a limit ordinal: Put $G_\alpha = \bigcup_{\beta \in [\omega,\alpha)} G_\beta$. Since $\omega \leq \beta < \gamma < \alpha$ implies $G_\beta \subseteq G_\gamma$, $G_\alpha$ is a function. Also, $Q(\alpha)$ is true: Clearly $G_\beta \subseteq G_\alpha$ for $\beta \in [\omega,\alpha)$. Consider any property $q = \langle z,\varphi \rangle$ where $z \subseteq \alpha$. There is $\beta \in [\omega,\alpha)$ such that $z \subseteq \beta$. By $Q(\beta)$, some $P \subseteq G_\alpha \restriction ((\omega \cup z) \times \omega)$ gives property $q$.

If $\alpha = \beta+1$: Form a sequence $\langle q_n \mid n \in \omega \rangle$ of the properties $q = \langle z,\varphi \rangle$ such that $z \subseteq \alpha$ and $\beta \in z$. By induction on $n$, we will find $P_n \supseteq P_n^o = \bigcup_{i<n} P_i$ such that $P_n \in \mathbb{P}\{\beta\}$ and there exists $P \subseteq G_\beta$ for which $P \cup P_n$ gives property $q_n$. Since we want the new part of the construction to yield elementary extensions of what is yielded from the previous construction, we will find the new by examining the old.

Consider $q_n = \langle z, \varphi(\vec{x}) \rangle$. Find 1-1 onto $g: \omega \cup z \to \omega \cup z-\{\beta\}$ such that $g \restriction (z-\{\beta\})$ is identity. Put $P^o = G_\beta \restriction \mathrm{dom}(gP_n^o)$. For symmetry (2), find $p: \mathrm{dom}(P^o) \to 2$ such that $pgP_n^o = P^o$. By $Q(\beta)$, there exists some $P^* \subseteq G_\beta \restriction ((\omega \cup z-\{\beta\}) \times \omega)$ such that $P^*$ gives the property $pgq_n$. By otherwise taking $P^o \cup P^* \subseteq G_\beta \restriction ((\omega \cup z-\{\beta\}) \times \omega)$, we can assume $P^o \subseteq P^*$. Put

$$P_n = ((pg)^{-1}P^*) \restriction (\{\beta\} \times \omega) \supseteq ((pg)^{-1}P^o) \restriction (\{\beta\} \times \omega) = P_n^o \restriction (\{\beta\} \times \omega) = P_n^o.$$

Note that $((pg)^{-1}P^*) \restriction ((z-\{\beta\}) \times \omega) \subseteq G_\beta$ since $\mathrm{dom}(p)$ is disjoint from $(z-\{\beta\}) \times \omega$ and $g$ is identity on $z-\{\beta\}$. For symmetry (2), find $p_1: (((\omega-z) \times \omega) \cap \mathrm{dom}((pg)^{-1}P^*)) \to 2$ such that

$p_1(((pg)^{-1}p^*) \restriction ((\omega - z) \times \omega)) \subseteq G_\beta$. Put $P = p_1(((pg)^{-1}p^*) \restriction ((\omega \cup z - \{\beta\}) \times \omega)) \subseteq G_\beta$. Then $P \cup P_n = p_1(pg)^{-1}p^*$ gives property $p_1(pg)^{-1}pgq_n = p_1q_n$. Since $p_1 \restriction Nz$ is the identity, $p_1q_n = \langle z, \varphi(p_1\vec{x})\rangle = q_n$.

Put $G_\alpha = G_\beta \cup \bigcup_{i<\omega} P_i$. The above construction guarantees $Q(\alpha)$. For the sublemma, put $G = G_{\aleph_1}$. QED Sublemma.

Consider any $N\omega \cup z$. By (1) of the sublemma, $G$ determines an ultrafilter $\mathcal{F} = \{[\![\varphi]\!]_{\omega\cup z} \mid \exists P \subseteq G(P \leq [\![\varphi]\!]_{\omega\cup z})\}$ on the Boolean algebra of values $[\![\varphi]\!]_{\omega\cup z}$. Then $\mathcal{F}$ determines relations $\circleddash$, $E$ on $N\omega \cup z$ by:

$$x \circleddash y \quad \text{iff} \quad [\![x = y]\!] \in \mathcal{F}$$

$$x\ E\ y \quad \text{iff} \quad [\![x \in y]\!] \in \mathcal{F}$$

Let $E$ interpret $\varepsilon$ and $\circleddash$ interpret the equality symbol. By (2), $[\![\forall v \psi(v)]\!]_{\omega\cup z} \in \mathcal{F}$ iff $\forall u \in N\omega \cup z([\![\psi(u)]\!]_{\omega\cup z} \in \mathcal{F})$. It follows (by induction on formulas) that, for all $\varphi$, $\langle N\omega \cup z, E, \circleddash\rangle \models \varphi$ iff $[\![\varphi]\!]_{\omega\cup z} \in \mathcal{F}$ iff $\exists P \subseteq G(P \leq [\![\varphi]\!]_{\omega\cup z})$. Thus $\langle N\omega \cup z, E, \circleddash\rangle$ satisfies all sentences which have Boolean value $\mathbb{1}$.

By (3), $x\,E\,y$ iff $\exists u \in \mathrm{dom}(y)(x \circleddash u \ \&\ u\ E\ y)$. By the above, $\langle N\omega \cup z, E, \circleddash\rangle$ is a model of the extensionality and equality axioms; further, $u \in \mathrm{dom}(y)$ is a well-founded relation of $u$, $y$. Thus, the function

$$H(y) = \{H(u) \mid u \in \mathrm{dom}(y) \text{ and } u\,E\,y\}$$

is well-defined such that: $x\,E\,y$ iff $H(x) \in H(y)$; $x \circleddash y$ iff $H(x) = H(y)$; and $H``N\omega \cup z$ is transitive. (The last sentence is essentially a result of Mostowski.)

Put $Mz = H``N\omega \cup z$.

Clearly, $z \subseteq z'$ implies that the $H$ defined for $N\omega \cup z'$ is an extension of the $H$ defined for $N\omega \cup z$. Thus, $z \subseteq z'$ implies

$Mz = H``N\omega \cup z \subseteq H``N\omega \cup z' = Mz'$.

Given $z \subseteq z'$, $\vec{x} \in N\omega \cup z$, we have

$$\langle Mz, \in\rangle \models \varphi(H(\vec{x})) \quad \text{iff} \quad \exists P \subseteq G(P \le [\![\varphi(\vec{x})]\!]_{\omega \cup z})$$

$$\text{iff} \quad \exists P \subseteq G(P \le [\![\varphi(\vec{x})]\!]_{\omega \cup z'}) \qquad \text{(used 4.7)}$$

$$\text{iff} \quad \langle Mz', \in\rangle \models \varphi(H(\vec{x})).$$

Thus, $Mz$ is an elementary submodel of $Mz'$.

Since $N\omega$ is a Boolean model of the admissibility axioms, $M0$ is admissible. It follows that each $Mz$ is admissible.

$Mz \cap Mz' = Mz \cap z'$: Suppose $t \in Mz \cap Mz'$. Then there are $x \in N\omega \cup z$ and $y \in N\omega \cup z'$ such that $H(x) = t = H(y)$. For some $P \subseteq G$, $P \le [\![x \equiv y]\!]$. By Corollary 4.13, there is $w \in N(\omega \cup (z \cap z'))$ such that $P \le [\![x = w]\!]$. Thus, $t = H(x) = H(w) \in Mz \cap z'$.

For each $\alpha \in \aleph_1$, define $a_\alpha = H(b_\alpha)$ where $b_\alpha$ comes from Lemma 4.11. Then $a_\alpha \in M\{\alpha\}$. Since $[\![b_\alpha \equiv b_\beta]\!] = \mathbb{0}$, $a_\alpha \ne a_\beta$.

$Mz \supseteq M$:

$$Mz \supseteq H``Nz \supseteq H``\{\check{x} \mid x \in M\} = M.$$

$o(Mz) = o(M)$: $o(M) \subseteq o(Mz)$ follows from the preceding; $o(M) \supseteq o(Mz)$ follows from observing that the rank of $H(x)$ is no greater than that of $x$.

## REFERENCES


1. H. Bachmann, Transfinite Zahlen, 2nd ed., Springer-Verlag, Berlin, 1967.

2. J. Barwise, *Infinitary logic and admissible sets*, Doctoral Dissertation, Stanford University, Stanford, Calif., 1967.

3. J. Barwise, *Implicit definability and compactness in infinitary languages*, in: The Syntax and Semantics of Infinitary Languages, Springer-Verlag, Berlin, 1968, 1-35.

4. J. Barwise, *Applications of strict* $\Pi_1^1$ *predicates to infinitary logic*, mimeographed, Yale, 1968-1969.

5. J.R. Buchi, *Die Boole'sche Partialordnung und die Paarung von Gefuegen*, Portugaliae Mathematica 7(1948), 119-190.

6. P.J. Cohen, *The independence of the continuum hypothesis, Parts I, II*, Proceedings of the National Academy of Sci. U.S.A. 50(1963), 1143-1148; 51(1964), 105-110.

7. K. Gödel, The Consistency of the Continuum Hypothesis, Princeton University Press, Princeton, N.J., 1940.

8. R.B. Jensen, Modelle der Mengenlehre, Springer-Verlag, Berlin, 1967.

9. R.B. Jensen and C.R. Karp, *Primitive recursive set functions*, in: Axiomatic Set Theory, part 1, American Mathematical Society, Providence, 1971, 143-176.

10. C.R. Karp, Languages with Expressions of Infinite Length, North-Holland, Amsterdam, 1964.

11. C.R. Karp, *Nonaxiomatizability results for infinitary systems*, Journal of Symbolic Logic 32(1967), 367-384.

12. C.R. Karp, *An algebraic proof of the Barwise compactness theorem*, in: The Syntax and Semantics of Infinitary Languages, Springer-Verlag, Berlin, 1968, 80-95.

13. G. Kreisel, *Model-theoretic invariants; applications to recursive and hyperarithmetic operations*, in: The Theory of Models, North-Holland, Amsterdam, 1965, 190-205.

14. G. Kreisel, *A survey of proof theory*, Journal of Symbolic Logic 33(1968), 321-388.

15. K. Kunen, *Implicit definability and infinitary languages*, Journal of Symbolic Logic 33(1968), 446-451.

16. A. Lévy, *The interdependence of certain consequences of the axiom of choice*, Fundamenta Mathematica 54(1964), 135-157.

17. A. Lévy, *Definability in axiomatic set theory I*, in: Proceedings of the 1964 International Congress for Logic, Methodology, and Philosophy of Science, North-Holland Publ. Co., Amsterdam, 1966, 127-151.

18. A. Lévy, *A hierarchy of formulas in set theory*, Memoirs of the American Mathematical Society, No. 57(1965).

19. A. Lévy and R.M. Solovay, *Measurable cardinals and the continuum hypothesis*, Israel Journal of Mathematics 5(1967), 234-238.

20. R. Platek, *Foundations of Recursion Theory*, Doctoral Dissertation, Stanford University, Stanford, Calif., 1966.

21. H. Rasiowa and R. Sikorski, The Mathematics of Metamathematics, Panstwowe Wydawnictwo Naukowe, Warszawa, 1963.

22. D. Scott, *Lectures on Boolean-valued models for set theory*, unpublished lecture notes of the U.C.L.A. Summer Institute on Set Theory, 1967.

23. D. Scott and R.M. Solovay, *Boolean-valued models of set theory*, to appear.

24. J.R. Shoenfield, *Unramified forcing*, in: Axiomatic Set Theory, I, American Mathematical Society, Providence, 1971, 357-382.

25. R. Sikorski, Boolean Algebras, 2nd ed., Academic Press, New York, 1964.

26. P.C. Suppes, Axiomatic Set Theory, Van Nostrand, Princeton, N.J., 1960.

Vol. 342: Algebraic K-Theory II, "Classical" Algebraic K-Theory, and Connections with Arithmetic. Edited by H. Bass. XV, 527 pages. 1973.

Vol. 343: Algebraic K-Theory III, Hermitian K-Theory and Geometric Applications. Edited by H. Bass. XV, 572 pages. 1973.

Vol. 344: A. S. Troelstra (Editor), Metamathematical Investigation of Intuitionistic Arithmetic and Analysis. XVII, 485 pages. 1973.

Vol. 345: Proceedings of a Conference on Operator Theory. Edited by P. A. Fillmore. VI, 228 pages. 1973.

Vol. 346: Fučík et al., Spectral Analysis of Nonlinear Operators. II, 287 pages. 1973.

Vol. 347: J. M. Boardman and R. M. Vogt, Homotopy Invariant Algebraic Structures on Topological Spaces. X, 257 pages. 1973.

Vol. 348: A. M. Mathai and R. K. Saxena, Generalized Hypergeometric Functions with Applications in Statistics and Physical Sciences. VII, 314 pages. 1973.

Vol. 349: Modular Functions of One Variable II. Edited by W. Kuyk and P. Deligne. V, 598 pages. 1973.

Vol. 350: Modular Functions of One Variable III. Edited by W. Kuyk and J.-P. Serre. V, 350 pages. 1973.

Vol. 351: H. Tachikawa, Quasi-Frobenius Rings and Generalizations. XI, 172 pages. 1973.

Vol. 352: J. D. Fay, Theta Functions on Riemann Surfaces. V, 137 pages. 1973.

Vol. 353: Proceedings of the Conference on Orders, Group Rings and Related Topics. Organized by J. S. Hsia, M. L. Madan and T. G. Ralley. X, 224 pages. 1973.

Vol. 354: K. J. Devlin, Aspects of Constructibility. XII, 240 pages. 1973.

Vol. 355: M. Sion, A Theory of Semigroup Valued Measures. V, 140 pages. 1973.

Vol. 356: W. L. J. van der Kallen, Infinitesimally Central-Extensions of Chevalley Groups. VII, 147 pages. 1973.

Vol. 357: W. Borho, P. Gabriel und R. Rentschler, Primideale in Einhüllenden auflösbarer Lie-Algebren. V, 182 Seiten. 1973.

Vol. 358: F. L. Williams, Tensor Products of Principal Series Representations. VI, 132 pages. 1973.

Vol. 359: U. Stammbach, Homology in Group Theory. VIII, 183 pages. 1973.

Vol. 360: W. J. Padgett and R. L. Taylor, Laws of Large Numbers for Normed Linear Spaces and Certain Fréchet Spaces. VI, 111 pages. 1973.

Vol. 361: J. W. Schutz, Foundations of Special Relativity: Kinematic Axioms for Minkowski Space Time. XX, 314 pages. 1973.

Vol. 362: Proceedings of the Conference on Numerical Solution of Ordinary Differential Equations. Edited by D. Bettis. VIII, 490 pages. 1974.

Vol. 363: Conference on the Numerical Solution of Differential Equations. Edited by G. A. Watson. IX, 221 pages. 1974.

Vol. 364: Proceedings on Infinite Dimensional Holomorphy. Edited by T. L. Hayden and T. J. Suffridge. VII, 212 pages. 1974.

Vol. 365: R. P. Gilbert, Constructive Methods for Elliptic Equations. VII, 397 pages. 1974.

Vol. 366: R. Steinberg, Conjugacy Classes in Algebraic Groups (Notes by V. V. Deodhar). VI, 159 pages. 1974.

Vol. 367: K. Langmann und W. Lütkebohmert, Cousinverteilungen und Fortsetzungssätze. VI, 151 Seiten. 1974.

Vol. 368: R. J. Milgram, Unstable Homotopy from the Stable Point of View. V, 109 pages. 1974.

Vol. 369: Victoria Symposium on Nonstandard Analysis. Edited by A. Hurd and P. Loeb. XVIII, 339 pages. 1974.

Vol. 370: B. Mazur and W. Messing, Universal Extensions and One Dimensional Crystalline Cohomology. VII, 134 pages. 1974.

Vol. 371: V. Poenaru, Analyse Différentielle. V, 228 pages. 1974.

Vol. 372: Proceedings of the Second International Conference on the Theory of Groups 1973. Edited by M. F. Newman. VII, 740 pages. 1974.

Vol. 373: A. E. R. Woodcock and T. Poston, A Geometrical Study of the Elementary Catastrophes. V, 257 pages. 1974.

Vol. 374: S. Yamamuro, Differential Calculus in Topological Linear Spaces. IV, 179 pages. 1974.

Vol. 375: Topology Conference 1973. Edited by R. F. Dickman Jr. and P. Fletcher. X, 283 pages. 1974.

Vol. 376: D. B. Osteyee and I. J. Good, Information, Weight of Evidence, the Singularity between Probability Measures and Signal Detection. XI, 156 pages. 1974.

Vol. 377: A. M. Fink, Almost Periodic Differential Equations VIII, 336 pages. 1974.

Vol. 378: TOPO 72 – General Topology and its Applications Proceedings 1972. Edited by R. Alò, R. W. Heath and J. Nagata XIV, 651 pages. 1974.

Vol. 379: A. Badrikian et S. Chevet, Mesures Cylindriques Espaces de Wiener et Fonctions Aléatoires Gaussiennes. X 383 pages. 1974.

Vol. 380: M. Petrich, Rings and Semigroups. VIII, 182 pages 1974.

Vol. 381: Séminaire de Probabilités VIII. Edité par P. A. Meyer IX, 354 pages. 1974.

Vol. 382: J. H. van Lint, Combinatorial Theory Seminar Eindhoven University of Technology. VI, 131 pages. 1974.

Vol. 383: Séminaire Bourbaki – vol. 1972/73. Exposés 418-435 IV, 334 pages. 1974.

Vol. 384: Functional Analysis and Applications, Proceeding 1972. Edited by L. Nachbin. V, 270 pages. 1974.

Vol. 385: J. Douglas Jr. and T. Dupont, Collocation Methods fo Parabolic Equations in a Single Space Variable (Based on C Piecewise-Polynomial Spaces). V, 147 pages. 1974.

Vol. 386: J. Tits, Buildings of Spherical Type and Finite BN Pairs. IX, 299 pages. 1974.

Vol. 387: C. P. Bruter, Eléments de la Théorie des Matroïdes V, 138 pages. 1974.

Vol. 388: R. L. Lipsman, Group Representations. X, 166 pages 1974.

Vol. 389: M.-A. Knus et M. Ojanguren, Théorie de la Descent et Algèbres d' Azumaya. IV, 163 pages. 1974.

Vol. 390: P. A. Meyer, P. Priouret et F. Spitzer, Ecole d'Eté d Probabilités de Saint-Flour III – 1973. Edité par A. Badrikia et P.-L. Hennequin. VIII, 189 pages. 1974.

Vol. 391: J. Gray, Formal Category Theory: Adjointness for 2 Categories. XII, 282 pages. 1974.

Vol. 392: Géométrie Différentielle, Colloque, Santiago c Compostela, Espagne 1972. Edité par E. Vidal. VI, 225 page 1974.

Vol. 393: G. Wassermann, Stability of Unfoldings. IX, 164 page 1974.

Vol. 394: W. M. Patterson 3rd, Iterative Methods for the Solutic of a Linear Operator Equation in Hilbert Space – A Surve III, 183 pages. 1974.

Vol. 395: Numerische Behandlung nichtlinearer Integrodiffere tial- und Differentialgleichungen. Tagung 1973. Herausgegebe von R. Ansorge und W. Törnig. VII, 313 Seiten. 1974.

Vol. 396: K. H. Hofmann, M. Mislove and A. Stralka, The Pontr agin Duality of Compact O-Dimensional Semilattices and Applications. XVI, 122 pages. 1974.

Vol. 397: T. Yamada, The Schur Subgroup of the Brauer Grou V, 159 pages. 1974.

Vol. 398: Théories de l'Information, Actes des Rencontres Marseille-Luminy, 1973. Edité par J. Kampé de Fériet et C. Pica XII, 201 pages. 1974.